Contents

Abstract

The British Pimplinae are revised, and illustrated keys are provided for their identification. For each species details of known distribution and biology in Britain are summarised. This treatment is supplemented by a check list, notes on techniques, notes on the systematics of Pimplinae, and a general review of the biology of the subfamily. One new genus and two new species are described and there are taxonomic and nomenclatural changes which affect a further 18 species.

Introduction

The subfamily Pimplinae is biologically the most diverse, and probably the most extensively studied, within the Ichneumonidae. In the British Isles there are 108 species, representing 37 genera. Worldwide the subfamily comprises about 77 genera. The majority of pimplines are ectoparasitoids of immature Lepidoptera, Coleoptera, Hymenoptera, Diptera or Arachnida, while some are pseudo-parasitoids of spider egg sacs, and one group are endoparasitoids in endopterygote pupae and prepupae. Several species are hyperparasitic, often facultatively, and a few are cleptoparasitic on other pimplines. Many of the genera include species of economic importance. Indeed, it is partly because many species are parasitoids of pests that so much (in comparison with other ichneumonids) is known about the biology of the group.

The Pimplinae is only a moderately large group but, because it includes several big, strikingly-coloured or common species, it is often numerically the best represented subfamily in general collections of Ichneumonidae. Such collections may be dominated by only three or four species, typically Pimplini that are common in suburban situations and also, but less frequently, *Rhyssa* and *Dolichomitus*, conspicuous on account of their large size and long ovipositors. This situation pertains throughout most of the world. In Britain, collections tend to be dominated by species of the genera *Pimpla* and *Itoplectis*. The numerous smaller, inconspicuous pimplines are less often collected and, although many appear to be quite widespread, usually few are present in museum insect collections. Malaise traps yield reasonable numbers but specialist collecting and rearing bring the best results. Many species thought to be 'rare' have proved to be widespread and locally common once their habits were discovered. Ecological specialisation coupled with the great ranges of hosts and biologies exhibited by the subfamily makes the Pimplinae one of the most attractive of all ichneumonid groups for experimental investigation.

The initial specimen base for this handbook project was the collection of the British Museum (Natural History) [abbreviated to BMNH]. The BMNH collection of Pimplinae was already relatively well organised and reflected the attention paid to it by J. F. Perkins, particularly during the early part of his career at the museum. In the immediate past much new material, often the product of modern, mass collecting techniques, has been added. Latterly the collection of the National Museums of Scotland [NMS] has increased both in size and in importance because of the accumulation of specimens with associated biological data, and it includes much reared pimpline material resulting from the present study. The other important collections of Ichneumonidae in Britain and Ireland have also been examined and have provided additional, invaluable data.

Members of the Pimplinae can be identified to subfamily level using the keys of Perkins (1959) or Townes (1969). However, neither of these is easy for the beginner. In Perkins' key pimplines run out in several places (couplets 22, 36, 37, 39 and 42) and his concept of the subfamily differed from that now accepted. It then included the Acaenitinae (couplet 24) and Xoridinae (couplet 22). Unfortunately pimplines do not have a simple combination of diagnostic features which can be used to separate them easily from other ichneumonids. Many have a recognisable general appearance (Figs 1, 159), but certain members of other subfamilies are superficially similar (and were, indeed, used to be classified as pimplines). The most common of these are females with long ovipositors belonging to genera such as *Lissonota* (Banchinae). The keys to subfamilies cited above and the following characters will help in some cases of doubt:

Tarsal claws never pectinate, but often with a basal lobe or tooth (see Figs 56, 60, 61).

Ovipositor always projecting obviously and its apex never with a dorsal notch, but sometimes with a nodus (see Fig. 9).

Propodeal carination, if reduced, never consisting of the posterior transverse carina alone.

Fore wing often with vein 3rs-m present and with the areolet shaped as in Figs 3, 12, 13, 25, 95, 96.

The British xoridines were dealt with by Gauld & Fitton (1981) and the acaenitines by Fitton (1981) and Shaw (1986).

Systematics

The subfamily is usually divided into seven tribes—Ephialtini, Polysphinctini, Rhyssini, Poemeniini, Pimplini, Delomeristini and Diacritini—but there is no general agreement on their exact limits (compare, for instance, Finlayson, 1967 and Townes, 1969). The Ephialtini is probably the most primitive group, and it is almost certainly a paraphyletic 'grade-group' from within which at least some of the other tribes have arisen. The Delomeristini appears to be a polyphyletic assemblage but it cannot be broken down in any obvious way, and its component genera cannot be related unambiguously to other tribes. The five remaining tribes seem to be natural, holophyletic groups, although the limits of one, the Polysphinctini, can be altered depending upon whether its definition is based on larval or adult characters (Townes, 1969; Gupta & Tikar, 1978).

At the species level many taxonomic problems remain to be solved. None of the published systematic studies of European Pimplinae can be described as a comprehensive revision. Thus the taxonomy of the British pimplines as set out in this handbook must be seen in the context of the comparable faunistic works dealing with other parts of the Palaearctic region and less than satisfactory studies of individual genera. Consequently, there is a need to be aware of the possible existence of additional species and to continue to question the integrity of the species recognised. This does not mean that we wish to avoid responsibility for taxonomic judgements made, but rather to emphasise that much remains to be done. And, lest it be forgotten, the Pimplinae has probably received more attention than most of the other large subfamilies of the Ichneumonidae! On the nomenclatural side the Pimplinae is relatively well served: with two recent catalogues (Oehlke, 1967; Aubert, 1969), built on the foundations laid by Perkins (1940; 1941; 1943), Townes & Townes (1960) and Townes, Momoi & Townes (1965). There are problems caused by dual usage of some generic names which confuse the uninitiated (see the notes under *Ephialtes, Pimpla* and *Apechthis*).

Morphological and biological variation within some of the 'morphospecies' that we have recognised suggests the existence of complexes of closely related biological species. Genera with particular problem areas include *Scambus, Liotryphon* and *Dolichomitus*. For example, when a 'species' has been reared in numbers from more than one host it is often possible to separate and characterise 'host races', which at first sight appear to be good biological species. However, when one considers the whole range of miscellaneous, non-reared material in collections these segregates frequently become impossible to sustain, raising questions as to their validity and/or the existence of even more biological species within the complex. Unfortunately, in ichneumonids there are no straightforward ways of addressing such difficulties, such as seem to be available in some insect groups, for instance through genitalic characters. Progress might best be made from an experimental approach in these cases, relating variation in host range, behaviour and physiology under laboratory conditions to an analysis of morphology involving specimens of known origin.

General biology

Although rivalled by the much larger subfamily Phygadeuontinae, the Pimplinae is remarkable among ichneumonid subfamilies for its wide range of host associations. This is no doubt a reflection of the generally unspecialised mode of development of the early stages seen in the primitive tribe, or ancestral 'grade-group', Ephialtini. Typically the host lives in a concealed situation and it is stung more or less to death at the time of oviposition, when the fully yolked egg(s) is laid externally to it. The conditions that the developing parasitoid larvae experience (essentially of feeding on more or less liquid carrion, from the outside) are thus physiologically relatively undemanding, and have little to do with limiting host ranges. The more specialised tribes Rhyssini and (as far as is known) Poemeniini develop in similar ways; and even in the tribe Pimplini, which does not entirely conform to this behaviour as the egg is deposited inside a lepidopterous pupa, the host is normally inactivated soon after attack. In the case of *Pimpla* species, at least, this is accomplished by the newly hatched parasitoid larva, which quickly migrates to the host's brain (Führer & Kilincer, 1972); although selective egg placement (Carton, 1978) and injected secretions from the female's accessory

5

glands (Führer, 1975; Osman, 1978) are also important in helping to overcome the host's defences. Such parasitoids, which allow the host practically no activity or development after attack, are termed idiobionts (Askew & Shaw, 1986): characteristically they have host ranges that may be relatively broad and are in any case focussed on ecological niches more than towards taxonomic groupings of their hosts. The only Pimplinae known to permit their hosts to continue to move freely, feed and grow after being parasitised are in the tribe Polysphinctini, all members of which develop as external parasitoids on more or less immature spiders. Parasitoids like these, whose hosts remain active for a significant time after being attacked, are termed koinobionts (Askew & Shaw, 1986) and tend to have further, taxonomic, restrictions on their host ranges in accordance with the greater demands of adapting to a living and physiologically active host.

The large overall diversity in the niches used by extant idiobiont pimplines supports the view that host ranges of idiobionts can be evolved particularly as a result of long-term ecological factors. In other words, the availability of hosts (i.e. a source of discoverable hosts that is dependable over evolutionary time) has been able to promote specialisation and diversification relatively easily, without inhibition from exacting host-mediated physiological constraints. This is in contrast with most koinobiont groups of parasitoids, in which the overall host range of the group normally has rather sharp taxonomic limits.

Despite the *potential* for relatively broad host ranges implicit in the basic biology of many pimplines, few of the species with adequately known biologies are really generalists. Instead a given species tends to have a host range that is easily characterisable, usually in terms of an ecological niche within which, however, quite an assortment of host species may be used if encountered. Examples of niche specialisation with great disparity in the actual hosts used are easy to see in species of *Scambus* (Ephialtini), but this pattern is perhaps most strikingly illustrated by relatively more specialised species such as *Rhyssa persuasoria* (Rhyssini) and *Perithous divinator* (Delomeristini). These have very narrow host ranges in the sense of being exclusively associated with, respectively, siricid woodwasps and stem-nesting aculeate Hymenoptera, and yet in each case they are known to develop also as facultative secondary parasitoids (Hanson, 1939; Danks, 1971).

Facultative secondary parasitism within lepidopterous pupae is also quite common in the tribe Pimplini, and some species of *Itoplectis* regularly attack fully exposed ichneumonid and braconid cocoons as well (pseudohyperparasitism: Shaw & Askew, 1976). Secondary parasitism and cleptoparasitism are poorly differentiated when either the host or its primary parasitoid can serve as food, but when the tendency for the original host to serve is coupled with a complete dependence on prior attack by another parasitoid species it is usual to refer to the dependent species as a cleptoparasite. For example, species of the cleptoparasitic genus *Pseudorhyssa* (Delomeristini) have weak slender ovipositors that are incapable of drilling through wood to reach the larva of their woodwasp hosts. For this they are dependent on the oviposition drills used by species of Rhyssini, so that the *Pseudorhyssa* egg is always deposited on an already parasitised host. The first instar larva of *Pseudorhyssa* is highly mobile and equipped with large powerful mandibles, enabling it to kill the rhyssine larva almost every time (Skinner & Thompson, 1960; Spradbery, 1969; 1970a). Cleptoparasitism and even the more ordinary forms of secondary parasitism are often only revealed by close study in the field, and this has not been given to many genera of Pimplinae. Some of our rarer species, in particular, may turn out to have complex requirements of this sort.

A number of studies on the general biology and behaviour of parasitic Hymenoptera have involved species of Pimplinae. This is partly because of the prominence and abundance of some species in the natural environment, coupled with hopes that the species could be manipulated as parasitoids of pest insects. But it is also because the rather undemanding and flexible nature of some species makes them good experimental animals. The basic requirements as larvae of most species also render them relatively easy to culture in the laboratory, sometimes on unnatural surrogate hosts (e.g. *Rhyssa* on *Apis* larvae: Spradbery, 1968; *Pimpla* on *Tenebrio* pupae: Sandlan, 1980) or even on completely artificial diets (e.g. *Itoplectis*: House, 1978). However, the general biology of the koinobiont Polysphinctini has hardly been reported in the literature at all, except that some species are recorded as using a venom that causes the host spider temporary paralysis, enabling the egg to be affixed (Nielsen, 1923; Cushman, 1926). The following brief review is aimed at giving an overview of the general biology of the idiobiont groups only. Where appropriate, more detailed biological information is given under tribal, generic and specific headings in the systematic part of the text.

Like most Hymenoptera, adults of many (probably all) species depend for their activity on an intake of carbohydrate, and feed on honeydew, nectar (Leius, 1960; Cole, 1967) and sometimes other plant secretions (Juillet, 1959). As is usual for synovigenic parasitoids, which produce a succession of fully yolked eggs but normally emerge as adults with relatively undeveloped ovaries, they also feed on individuals of their host groups. These are killed and usually so severely mutilated by this process that they are rarely used for oviposition subsequently (Jackson, 1937; Juillet, 1959; Leius, 1961a; Cole, 1967; Sandlan, 1979a). Both churning with the ovipositor and enlarging the resulting wound with the mouthparts (Jackson, 1937; Leius, 1961b; Sandlan, 1979a) can contribute to the flow of haemolymph that is imbibed by the female parasitoid. Sometimes undersized or otherwise unsuitable 'host' individuals (including species outside the realised host range) are used for host feeding (Cole, 1967; Sandlan, 1979a) and in one North American species of *Itoplectis* regular mutilation and host feeding attacks on pupae of the introduced moth *Lymantria dispar* (Linnaeus) have been found to cause up to 200 times as much mortality as results from actual parasitism (Campbell, 1963). The several laboratory studies correlating adult feeding on either hosts or carbohydrate from plant sources with increases in longevity, fecundity and accessory gland function indicate that undertaking both kinds of feeding is essentially obligatory in many pimplines (Leius, 1961a, b; Sandlan, 1979a; Osman, 1978). This is known to apply to Pimplini and some genera of Ephialtini that attack weakly concealed hosts: further observation is needed on other groups, and especially those that attack hosts deeply concealed in wood, to ascertain the generality of host-feeding in the subfamily.

Adult pimplines find their hosts by a variety of successive environmental cues. Odours from the general substrate supporting the host (Thorpe & Caudle, 1938) or from symbionts more definitely indicating the presence of a host (Spradbery & Kirk, 1978), as well as odours and contact chemicals stemming from the host itself (Sandlan, 1980) and even vision (Sandlan, 1980) may all play a part. Orientation towards, and recognition of, the searching niche—'host habitat finding'—by parasitic Hymenoptera is usually recognised as clearly mechanistically distinct from the discovery and recognition of an actual host individual (Doutt, 1964; Vinson, 1976; 1981; see also Weseloh, 1981 and Arthur, 1981). Some pimplines are known to be capable of simple associative learning (Arthur, 1966; Wardle & Borden, 1985), enabling them to concentrate their efforts on the most productive sections of the environment, and this suggests a mechanism whereby the realised host range of a particular species might differ radically from place to place. As is usual for idiobionts, many pimplines have adult flight periods that correspond to peaks of potential host availability rather than being attuned to a particular host species. Bivoltine pimplines often have essentially different realised hosts at different times of the year (Cole, 1967), although some of the more niche-specialised ones may make double use of a single host generation (Danks, 1971; see also Hanson, 1939).

Most pimplines are typical haplo-diploid Hymenoptera, in which unfertilised eggs give rise to male progeny and fertilised eggs result in females. In common with many idiobionts, mated females of several pimpline genera are known to be able to select the sex of their progeny by controlling the access of stored sperm to the egg as it passes down the oviduct; a process that is detectable by direct observation in at least one species (Cole, 1981). This enables them to use the larger host individuals preferentially for female progeny (Jackson, 1937; Aubert, 1959; Arthur & Wylie, 1959; Sandlan, 1979b), even in some cases when the size of the concealed host cannot be directly assessed visually (e.g. Askew & Shaw, 1986). When a range of suitable hosts is freely available it often happens in the field that the two sexes develop on substantially different groups of hosts (Kishi, 1970; Askew & Shaw, 1986).

Much remains to be discovered about the larval adaptations of not only pimplines but of parasitic Hymenoptera as a whole. However, various interesting scraps of information on the larval behaviour of particular species are reported in the literature, including the migration of the young larva of *Pimpla* to the brain of its host (Führer & Kilincer, 1972), the adaptations for fighting in the first instar of the cleptoparasitic *Pseudorhyssa* (Spradbery, 1969; 1970; Skinner & Thompson, 1960), the consuming of the host's skin after all the juices have been sucked out by *Perithous* (Danks, 1971), and Nielsen's (1923 *et seq.*) various observations on Polysphinctini and the Ephialtini most closely related to them. The larva of *Pimpla* produces a biochemically complex anal secretion that has antibiotic properties and presumably helps to keep the wounded host aseptic while it is being consumed (Führer & Willers, 1986). It has been shown that there are five larval instars in *Pimpla* (Rojas-Rousse & Benoit, 1977), the middle three being very similar and hard to distinguish, and this seems likely to be the case for all Pimplinae.

Cocoon formation is highly specialised only in the koinobiont tribe Polysphinctini. Some polysphinctines construct their cocoons within the host spider's web, in which case the cocoon is usually densely spun, spindle-shaped and sometimes angularly ribbed, no doubt resembling a well-wrapped dead fly or moth of little interest to its potential enemies. Other polysphinctines finally kill their hosts in retreats spun or selected by the spider, and the cocoon is then usually much frailer, expanded and loosely spun. Most polysphinctine cocoons have a clearly visible caudal opening through which the meconium (the accumulated larval faecal material) is voided. Elsewhere in the Pimplinae, however, cocoon formation appears to be rather generalised and is well developed only when the host or its substrate fails to provide a really secure or concealed pupation site, particularly if the pimpline passes the winter as a prepupa. Thus the most substantial cocoons are seen in pimpline genera which attack and overwinter in frail structures, like the exposed cocoons of Lepidoptera that are attacked by *Iseropus*. On the other hand, genera attacking hosts living deeply concealed in woody tissues or the hard pupae of Lepidoptera seem often to spin only enough silk to restrict the involuntary movement of the pupa, or sometimes no detectable silk at all. Another function of a cocoon is to isolate the occupant from micro-organisms, and some species inhabiting plant tissues that are decaying, or are likely to decay, appear to make relatively strong cocoons. The cocoons of most genera, and some entire tribes, remain unstudied, however. In common with other Hymenoptera, eclosion from the pupa precedes emergence from the cocoon by two or three days, during which the initially teneral adult parasitoid hardens so that it is fully protected when it finally bites its way out.

The majority of species of Polysphinctini overwinter as immature larvae on their fully functioning spider hosts, but overwintering as an early instar larva is far less accessible to the idiobiont tribes. However, *Tromatobia* females do not kill the spider eggs in the sacs they attack, and some *Tromatobia* species can therefore overwinter as small larvae feeding on fresh diapausing spider eggs. Otherwise the most usual method of overwintering is as a prepupa; though females of a few species overwinter as adults, carrying with them the male's contribution of stored sperm (e.g. *Itoplectis maculator*: Cole, 1967; possibly *Pimpla turionellae* in part: Jackson, 1937; *Scambus pomorum* and *Exeristes ruficollis*: Hancock, 1925). Adults of *I. maculator* are also known to aestivate (Cole, 1967). The influences of the host species and of environmental conditions on the onset of winter diapause by the prepupae of *Pimpla hypochondriaca* have been investigated by Claret & Carton (1975) and Claret (1973; 1978) respectively.

Many of the pimpline species that are known to overwinter as adults have a pronounced reddish coloration, which may be more cryptic than black in the winter months. Another, even clearer, ecological correlate of reddish body coloration is an association with reed beds and fens: this is easily seen in pimplines, and in fact it extends through most groups of parasitic Hymenoptera, but it has not been satisfactorily explained.

Techniques

Although particular mention of Pimplinae is sometimes appropriate, this section applies very generally to Ichneumonoidea as a whole.

COLLECTING. Traditional collecting methods involving simple hand-netting, supplemented by sweeping, will produce reasonable numbers of pimplines and other ichneumonoids, especially in warm overcast weather. However, trapping, using Malaise traps, is probably the best means of obtaining large, general samples from most habitats. The catch from even one Malaise trap will astonish those used only to collecting by hand. Also, the traps operate continuously and can be left unattended, freeing the collector to concentrate on hand-netting, sweeping, searching for hosts for subsequent rearing of their parasitoids, or making behavioural observations. Two of the most productive habitats for the less common pimplines in particular are reed beds with carr, and long established woodland with overmature trees.

Malaise traps are tent-like structures; the invention of the Swedish hymenopterist René Malaise. They are made from fine mesh netting which blocks the passage of flying insects. The top part of the trap leads to a collecting chamber which has a removable receptacle for the preservative-cum-killing agent. Siting of traps is important. The best sites are areas of high flight activity, such as gaps in hedges and the boundaries between different vegetation types. Detailed plans and instructions for making a trap were given by Townes (1972). Commercially made traps are available from Marris House Nets, 54 Richmond Park Avenue, Bournemouth, BH8 9DR. The best preservative/killing agent for use in Malaise traps is 95% ethanol (neat industrial methylated spirit). Isopropanol can be substituted and has the advantage that it can be obtained without an Excise certificate.

In warm, still weather specimens of larger species are best hand-netted individually. A net for such use must be fairly small and manoeuvrable. The diameter of the net ring and the depth of the bag are matters of personal choice, but the material from which net bags are made must be fine and open enough to allow the free passage of air. Because they are often thrust into vegetation, one must be prepared to replace the bags fairly frequently. Many workers prefer an all-purpose net, such as a kite net, for hand netting and more general sweeping. Sweeping is especially suited to collecting smaller species. Noyes (1982) gave details of a sweep-net specially developed for chalcids, but suitable for ichneumonids.

Other collecting methods using, for example, yellow trays or pitfall traps can produce worthwhile results in some circumstances. In winter hibernating females of some genera can be sought, for example by beating evergreen trees, dissecting grass tussocks and birds' nests, or prizing loose bark from dead wood. Several pimplines are known to overwinter as adults, and it is possible that more will be found to do so.

PRESERVATION. Killing agents for reared or manually collected specimens are a matter of personal preference, and all of the usual ones are suitable. However, best of all is to leave the ichneumonid to die naturally in a clean, roomy tube. This usually leaves it well groomed, almost fat-free, and relaxed enough to be easily mounted straight away. If it cannot be mounted within a few hours of death, the insect can be allowed to dry out completely (but only a large, or preferably ventilated, container will guard against mould formation), but dry specimens will always need to be relaxed at a later date to be mounted well. Relaxation takes about 12–24 hours: the best relaxant is the moisture emanating from the chopped up young leaves of cherry laurel (*Prunus laurocerasus* Linnaeus), which can be prepared in mid-summer and stored indefinitely, but the essential conditions of high humidity in a mould-free and air-tight container can be obtained quite successfully with damp, clean cotton-wool. While the above, and especially mounting the specimen when it is freshly dead, probably produces the very best results, it is relatively demanding and often impractical. In addition, the dry storage of unmounted Ichneumonidae places them at considerable risk, the legs and antennae being not only easily broken, but sometimes also important for identification. It can therefore be simpler to use as a killing jar a tube partly filled with 95% alcohol, which can then receive a label (in pencil or drawing ink, not biro) to become a safe and practical means for medium-term storage. This form of storage is really the only practical way of dealing with large unsorted samples of ichneumonids that cannot be mounted straight away. Kept in a cool dark place, such samples will last without too much deterioration for several years. The most noticeable effect of long term storage in alcohol is the loss of some colour. Details of how such bulk samples can be sent through the post were given by Mason (1974).

The most convenient way of finally storing and handling individual adult ichneumonids for identification is as traditional dry, pinned preparations. If specimens are to be a source of reference, their structural details must be readily appreciated, not hidden by glue, cardboard or other parts of the anatomy. Ease of safe handling is of paramount importance. Inefficiently mounted specimens are particularly liable to breakage in groups such as parasitic Hymenoptera which usually have to be viewed from several successive angles to be identified. The handling pin for all mounting methods should be continental-length (38 mm) stainless steel, no thinner than size 1 or it may bend or 'twang', and it should never be headless. Direct pinning is much the best for large and medium sized species and is satisfactory for even quite small specimens, though for these staging with a micropin to polyporus or Plastazote is also practical, provided the stage is not so long that it may swing about the carrier pin. Care is needed in placing the pin: through the mesothorax, to the right of the centre line and clear of the back of the fore coxae.

Small to medium sized specimens can be glued to the side of the pin. The glue most commonly used for this is white shellac gel, which is white shellac in alcohol. White shellac does not have all the waxes removed and thus is not brittle when dry. Unfortunately it is difficult to obtain in Britain, and the only source of supply known to us is John Myland Ltd, 80 Norwood High Street, London, SE27 9NW. Other glues might be suitable for this method but have not been tried by us.

For both direct pinning and glueing to the pin shaft, the insect should be positioned about two-thirds of the way up the shaft of a continental pin. This leaves room above it for handling and below it for data and determination labels, and ensures good all-round visibility under easy control. The antennae should be positioned well clear of the pin head, but there is no need at all to 'set' the specimen like a lepidopteran: indeed, many important characters at the side of the thorax may be obscured if this is done. Better is to pull the legs gently downwards and outwards if they are hunched to the body, and allow the wings to fold more or less upwards above the thorax.

Card pointing used to be recommended for mounting smaller parasitic Hymenoptera but it is not as popular as it once was, as it is easy to do badly but difficult to do well. Water-soluble glues such as Seccotine are best, in case of the need to remount.

To dry mount specimens from alcohol of any strength they should first be swirled with a little clean 95% alcohol, then removed and laid flat on absorbent tissue paper until damp-dry. They will be quite brittle after prolonged storage in 95% alcohol (less so from 75%), but minimal rearrangement of appendages may be possible, and if necessary the wings can be uncrumpled with a pin as they dry. The warmth of a table lamp can be useful in quickening the drying process, though at the risk of making antennae and legs more brittle. Indeed, it is advisable to mount specimens rather soon after their surface has dried to minimise damage.

REARING. Rearing parasitoids from their hosts is undoubtedly the method that will add most to our knowledge of their general biology, habitat preferences and host associations. For the practical methods appropriate to handling live host material it is best to consult publications on the host groups concerned and to obtain advice from specialists. As for the reared parasitoids there are a number of things to be taken into account. If possible, hosts should be reared individually. Great care needs to be taken to prevent the introduction into rearing cages of additional, potentially parasitized hosts with food plants, etc. The resulting erroneous associations may remain undetected and as sources of confusion for a very long time.

It is obviously important to record more of the information relating to reared material than is normal for field-collected adults: a single, unqualified date is completely meaningless, as it could be either a date of collection or a date of emergence. In our experience about half of all entomologists who give single dates for reared material use it for collection and the other half for emergence, yet the confusion is immediately cleared by the qualifying use of 'coll.' or 'em.' before the date if for some reason only one can be given. Ideal data would include name and stage of host; date and place of collection; date of adult parasitoid emergence (if this happens after a significant period indoors, record 'indoors' against the emergence date to indicate that it may not reflect emergence in the wild); and details and dates of intermediate events in the development of host and parasitoid. Also of importance are details of brood size of gregarious parasitoids; and, particularly, a clear indication of the ecological situation or micro-habitat from which the parasitized material was collected. If quantitative data are available they should also be noted. It may not be possible to give the name of the host with certainty, and for all such cases it is best to express the maximum doubt. This is extremely important for substrate rearings: for example, 'ex standing dead *Pinus sylvestris* with *Pissodes pini*, *Acanthocinus aedilis* and *Rhagium inquisitor*' is fully informative and much more reliable than making a guess on the basis of just the commonest possible host present. Indeed, if the exact host specimen is not absolutely certain and its remains are not recovered, it is best to record a reared parasitoid as, for example, 'ex *Calamagrostis epigejos* stems with *Calameuta filiformis*' even if this sawfly was in abundance and the only other insect reared from the stems collected. Often the host remains *will* be recoverable, however, and at least a family identification may then become certain but, even if 'Lepidoptera pupa' is all that can be given, the macro- and micro-habitat information will still be valuable and indeed may be crucial from the parasitoid's point of view. This is because most parasitoids (and this is especially true of many Pimplinae) tend to have a 'host range' that is at least partly niche-oriented, rather than being a fixed, limited set of host species. If several identical hosts are collected at the same time it is nevertheless important that they are kept singly if possible, so that each parasitized host can be unambiguously associated with the parasitoid(s) that emerge; and these host remains should always be individually preserved along with the respective adult parasitoids and their cocoons.

Preservation of host remains and cocoons is vital as it enables checks to be made on host identifications, in cases of doubt, and on some life history details, such as the occurrence of secondary parasitism. It also provides for a source of ichneumonid larval skins that are identifiable from the associated adults. It is therefore very important to fit the right cocoon to the right adult parasitoid (or to indicate doubt such as 'cocoons mixed'): another reason for trying to conduct individual rearings (though a brood of parasitoids that developed gregariously with respect to a single host should not be split up).

Adult parasitoids should not be killed straight after they have emerged: even if the starvation method is not employed, it is best to leave them alive for a few days. Host remains and cocoons should be allowed to dry out completely, then preserved dry in a gelatine capsule. This is (preferably) impaled on the pin on which the adult is mounted or, if there is not room, it can be put on a separate pin bearing identical labels and some other means of associating it with the individual adult to which it relates. If the material is not enclosed in a capsule there is a risk of loss of loose cast skins, etc. through the emergence hole left by the adult parasitoid.

Terminology and notes on the keys

MORPHOLOGY. Most of the terms used for morphological features are explained by means of annotated figures (Figs 1–10). A full, general account of morphology was given by Richards (1977) in the introductory Handbook to the Hymenoptera, and our terminology follows his in all but a few clearly indicated instances. Many of the characters used in the keys are illustrated and the relevant figures referred to directly. Terminology for cuticular microsculpture is as described and illustrated by Eady (1968). A selection of Eady's figures are reproduced as an appendix to this handbook (p. 98). The relative longitudinal positions of the valves of the ovipositor are not fixed and although most of those illustrated have the upper valve extending a little beyond the lower valves (as in Fig. 9) this is not always the case in preserved specimens.

MEASUREMENTS. Fore wing length, measured from the apex of the tegula to the apex of the wing, is used as an indication of overall size. This is easily and accurately measured and is much more consistent than overall body length. The ranges given cover both sexes. In general males are smaller than females, but for the purposes of this handbook separate measurements would be as misleading as they might be helpful. Unless otherwise noted, statements about size and shape of individual parts, such as gastral tergites and tarsal segments, are based on length measured medially and maximum width (as shown in Figs 7, 8). The scales given with figures all represent 0·5 mm and are intended as a guide to size, but it should be remembered that there is often very large variation within species and genera.

The relative length of the ovipositor is important. As a measure of it we use the *ovipositor–hind tibia index* which is the length of the ovipositor (and its sheath) projecting beyond the apex of the gaster (see Fig. 1) divided by the length of the hind tibia. Although, often, the point on the ovipositor which coincides with the apex of the gaster will have to be estimated (see Fig. 10), this exposed portion can be measured more accurately than the entire ovipositor.

SEX. Adult pimplines are easily sexed. In all British species the ovipositor and its sheath project conspicuously beyond the apex of the female gaster (see Figs 1, 10). There are no males in which the form of the genitalia (see Fig. 8) will lead to confusion with females.

KEYS. Where possible males and females are keyed together, although sex (see note above) is used as a key character at various points. However, for some genera we have not been able to construct a key to males. In such cases we have given some characters which will aid identification of males, especially those associated with females, under individual species.

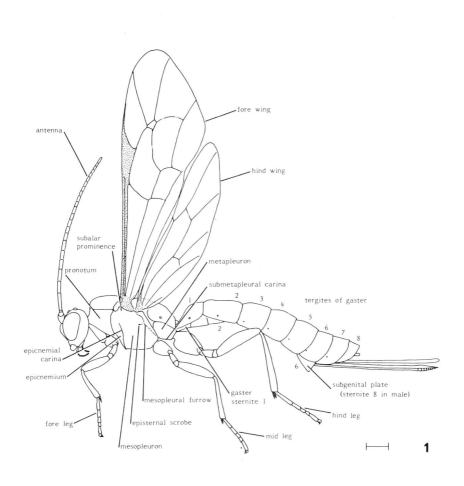

Fig. 1. Whole insect, lateral view, *Endromopoda detrita* female. Scale line represents 0·5 mm.

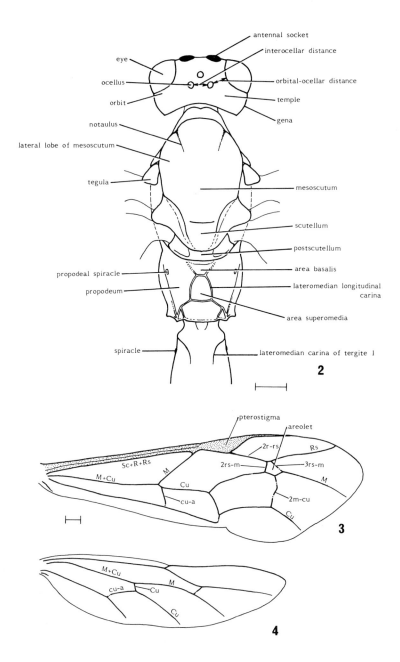

Figs 2–4. 2, head, thorax and propodeum, dorsal view, *Delomerista mandibularis*. 3, fore wing, *Pimpla hypochondriaca*. 4, hind wing, *Pimpla hypochondriaca*. Scale lines represent 0·5 mm.

14

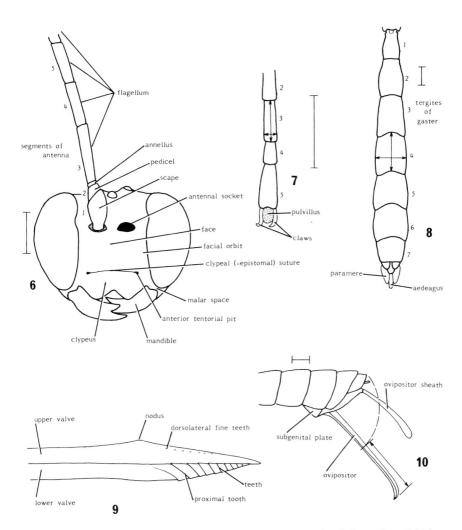
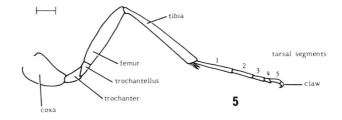

Figs 5–10. 5, left hind leg, lateral view, *Liotryphon crassisetus*. 6, head, front view, *Ephialtes manifestator*. 7, apex of hind tarsus, dorsal view, *Tromatobia oculatoria*. 8, gaster, dorsal view, *Rhyssella approximator* male. 9, apex of ovipositor, lateral view, diagrammatic. 10, apex of gaster, lateral view, *Apechthis quadridentatus*, to show method of estimating projecting length of ovipositor. All scales lines represent 0·5 mm.

Presentation of records

The notes on individual species are not given in the keys, but are listed separately in alphabetical order following the key to species for each genus. The extended nature of these notes, particularly in relation to host associations, is the reason for this change from the usual *Handbook* arrangement.

Published host records must, alas, always be treated with considerable caution and even downright suspicion. There are several reasons for this in addition to the obvious risk that the adult reared parasitoid may very well have been misidentified. One is that the *host* may have been misidentified: with the best will in the world it is surprisingly easy to overlook extraneous specimens of other potential host species during rearing programmes, especially if they are already dead through having been parasitized, and this risk is even greater when parasitoids are being reared from large, solid substrates such as logs of wood, in which the isolation of each potential host is impractical. Another problem in interpreting even genuine rearing records is that they usually lack any relative weighting. Unless rearing records are expressed in a quantitative way the regular host associations will not emerge as grains of truth, made dependable by repetition, from the chaff of misidentifications, errors of observation, and irregular occurrences. Rearing records of specimens that we have been able to examine are presented here in a quantitative way: these include host data taken from labels on museum specimens, but also a considerable number of specimens recently reared under carefully controlled conditions. The many rearing records of a rather vague nature— for example, substrate rearings or those from hosts identified only to Order or to family—in general have not been listed among the detailed rearing records, but they have in all cases been taken into account in the summaries of host range given for the species concerned. Records of pseudohyperparasitism are given in the form: primary parasitoid (identified to genus if possible)/its host (to genus if possible), and the accompanying family placements refer only to the primary parasitoids concerned. In general literature records are referred to only when there is little or no other source of biological information and, unless otherwise indicated, rearing records all refer to British specimens.

Similarly, our assessments of rarity and summaries of distribution and flight periods are based almost entirely on data from specimens which we have examined. Rarity is expressed as 'rare', 'uncommon' or 'common', sometimes with a qualification. For the rarest species, where we have seen material from fewer than 10 localities, the actual number of specimens is given in the form 'x specimens from y localities'. Macro-habitat information is given for some species. Detailed distributions of species within the British Isles are not given, but information is summarised using geographical terms and current names of administrative units (counties in England, Wales and Ireland and districts in Scotland). Flight periods are given using Roman numerals for months when adults have been collected (or reared if the information is of value). Some of the species referred to as 'bivoltine' will undoubtedly have at least partial third generations under favourable circumstances.

Names of hosts follow the most recent British check lists: of Lepidoptera, Diptera, Coleoptera, Hymenoptera, and spiders (Kloet & Hincks, 1972; 1976; 1977; 1978; Merrett *et al.*, 1985; respectively). Exceptions are the lepidopterous families Sesiidae and Choreutidae, for which we follow Heath & Emmet (1985).

Check list

The check list below supersedes the one published by Fitton (*in* Kloet & Hincks, 1978). Fifteen species (including two unnamed species, of *Endromopoda* and *Scambus*) are added and seven (including *Tromatobia rufipleura* (Bignell) which is no longer considered to be distinct from *T. ovivora* (Boheman)) are deleted. A sixteenth addition, *Zatypota albicoxa* (Walker), has already been published (Hudson, 1985). Taxonomic and nomenclatural changes affect another fourteen species and are shown in the list. They are explained at the appropriate points in the text. One of these changes unfortunately affects the common species well known as *Pimpla instigator*.

The named species added to the British List in this handbook are *Scambus annulatus* (Kiss), *S. cincticarpus* (Kriechbaumer), *S. foliae* (Cushman), *Endromopoda nigricoxis* (Ulbricht), *Tromatobia forsiusi* (Hellén), *Poemenia collaris* (Haupt), *Poemenia notata* Holmgren, *Itoplectis aterrima* Jussila, *I. insignis* Perkins, *Sinarachna nigricornis* (Holmgren), *Delomerista pfankuchi* (Brauns), *Polysphincta nielseni* Roman, *Polysphincta vexator* sp. nov. and *Pimpla wilchristi* sp. nov.

All the British specimens examined purporting to be the following species have turned out to be misidentified. They are therefore deleted from the British List and are not included in the keys to species.

Schizopyga varipes Holmgren. All the specimens proved to be *S. circulator* (Panzer).

Oxyrrhexis carbonator (Gravenhorst). This name has been widely misapplied in Europe and the only British specimen seen was a damaged *Zaglyptus multicolor* (Gravenhorst).

Perithous albicinctus (Gravenhorst). Morley (1908: 45) recorded this species as British because he incorrectly synonymized both *Ephialtes albicinctus* Desvignes and *Ichneumon annulatorius* Fabricius with it. *E. albicinctus* Desvignes (with its junior objective synonym *Ephialtes desvignesii* Marshall) is a junior synonym of *Delomerista mandibularis* (Gravenhorst) (Fitton *in* Kloet & Hincks, 1978: 14; Gupta, 1982a: 3, 25). *I. annulatorius* Fabricius is a North American species (Perkins, 1952: 67).

Neoxorides nitens (Gravenhorst). The specimen in the BMNH collection which probably led to the inclusion of this species in the British List (Morley, 1908: 16) is *Podoschistus scutellaris* (Desvignes).

Sinarachna anomala (Holmgren). We have seen no British material of this species. It was recorded from England by Oehlke (1967: 25), perhaps by mistake.

Scambus signatus (Pfeffer). This species was introduced as British by Perkins (1943). Our reasons for excluding it from the British List are discussed under *S. annulatus* (Kiss) (p. 50).

ICHNEUMONIDAE
PIMPLINAE
EPHIALTINAE

EPHIALTINI
PIMPLINI sensu Townes

EPHIALTES Gravenhorst, 1829
   PIMPLA sensu Curtis, 1828 suppressed
manifestator (Linnaeus, 1758)

DOLICHOMITUS Smith, 1877
agnoscendus (Roman, 1939)
diversicostae (Perkins, 1943)
imperator (Kriechbaumer, 1854)
mesocentrus (Gravenhorst, 1829)
messor (Gravenhorst, 1829)
populneus (Ratzeburg, 1848)
pterelas (Say, 1829)
terebrans (Ratzeburg, 1844)
tuberculatus (Geoffroy *in* Fourcroy, 1785)

TOWNESIA Ozols, 1962
tenuiventris (Holmgren, 1860)
antefurcalis (Thomson, 1877)

PARAPERITHOUS Haupt, 1954
gnathaulax (Thomson, 1877)

LIOTRYPHON Ashmead, 1900
caudatus (Ratzeburg, 1848)
crassisetus (Thomson, 1877)
albispiculus (Morley, 1908)
punctulatus (Ratzeburg, 1848)
ruficollis (Desvignes, 1856)
ascaniae (Rudow, 1883)
strobilellae (Linnaeus, 1758)

EXERISTES Foerster, 1869
EREMOCHILA Foerster, 1869
ruficollis (Gravenhorst, 1829)

AFREPHIALTES Benoit, 1953
cicatricosa (Ratzeburg, 1848)

FREDEGUNDA Fitton, Shaw & Gauld **gen. nov.**
diluta (Ratzeburg, 1852) **comb. nov.**

ENDROMOPODA Hellén, 1939 **stat. nov.**
arundinator (Fabricius, 1804) **comb. nov.**
melanopyga (Gravenhorst, 1829)
culpator (Morley, 1914)
detrita (Holmgren, 1860) **comb. nov.**
brunneus (Brischke, 1880)
nigricoxis (Ulbricht, 1910) **comb. nov.**
nitida (Brauns, 1898) **comb. nov.**
?*deplanata* (Morley, 1908)
arundinator misident.
phragmitidis (Perkins, 1957) **comb. nov.**
species *A*

SCAMBUS Hartig, 1838
EPIURUS Foerster, 1869
ATELEOPHADNUS Cameron, 1905
annulatus (Kiss, 1924)
?*signatus* (Pfeffer, 1913)
brevicornis (Gravenhorst, 1829)
anomalus (Morley, 1906)
buolianae (Hartig, 1838)
calobatus (Gravenhorst, 1829)
cincticarpus (Kriechbaumer, 1895) **stat. rev.**
elegans (Woldstedt, 1876)
ulicicida (Morley, 1911)
eucosmidarum (Perkins, 1957)
foliae (Cushman, 1938)
nigricans (Thomson, 1877)
similis (Bridgman, 1884)
robusta (Morley, 1908)
planatus (Hartig, 1838)
nucum (Ratzeburg, 1844)
pomorum (Ratzeburg, 1848)
sagax Hartig, 1838

linearis (Ratzeburg, 1844)
vesicarius (Ratzeburg, 1844)
 gallicola (Morley, 1908)
species *A*

ACROPIMPLA Townes *in* Townes & Townes, 1960
didyma (Gravenhorst, 1829)

ISEROPUS Foerster, 1869
stercorator (Fabricius, 1793)

GREGOPIMPLA Momoi *in* Townes, Momoi & Townes, 1965
inquisitor (Scopoli, 1763)

TROMATOBIA Foerster, 1869
forsiusi (Hellén, 1915)
oculatoria (Fabricius, 1798)
ornata (Gravenhorst, 1829)
 concors (Kriechbaumer, 1890)
ovivora (Boheman, 1821)
 rufipleura (Bignell, 1889) **syn. nov.**
variabilis (Holmgren, 1856)
 abdominalis (Brullé, 1846) homonym
 epeirae (Bignell, 1893)
 hibernica (Morley, 1908)

ZAGLYPTUS Foerster, 1869
multicolor (Gravenhorst, 1829)
varipes (Gravenhorst, 1829)

CLISTOPYGA Gravenhorst, 1829
incitator (Fabricius, 1793)
 haemorrhoidalis Gravenhorst, 1829
rufator Holmgren, 1856
sauberi Brauns, 1898
 ?*canadensis* Provancher, 1880

POLYSPHINCTINI

DREISBACHIA Townes, 1962
pictifrons (Thomson, 1877)
 bridgmanii (Bignell, 1894)

SCHIZOPYGA Gravenhorst, 1829
circulator (Panzer, 1801)
frigida Cresson, 1870
 atra Kriechbaumer, 1887
podagrica Gravenhorst, 1829

PIOGASTER Perkins, 1958
albina Perkins, 1958
punctulata Perkins, 1958

POLYSPHINCTA Gravenhorst, 1829
boops Tschek, 1868
nielseni Roman, 1923
rufipes Gravenhorst, 1829
tuberosa Gravenhorst, 1829
vexator Fitton, Shaw & Gauld **sp. nov.**

ACRODACTYLA Haliday, 1838
BARYPUS Halday *in* Curtis, 1837 homonym
COLPOMERIA Holmgren, 1859
SYMPHYLUS Foerster, 1871
POLEMOPHTHORUS Schulz, 1911
degener (Haliday, 1838)
madida (Haliday, 1838)
quadrisculpta (Gravenhorst, 1820)

SINARACHNA Townes *in* Townes & Townes, 1960
nigricornis (Holmgren, 1860)
pallipes (Holmgren, 1860)

ZATYPOTA Foerster, 1869
albicoxa (Walker, 1874)
eximia (Schmiedeknecht, 1907)
bohemani (Holmgren, 1860)
discolor (Holmgren, 1860)
  thoracica (Brischke, 1864)
  percontatoria: Aubert, 1969, 1970
percontatoria (Müller, 1776)
  phoenicea (Haliday, 1838)
  gracilis (Holmgren, 1860)
  scutellaris (Holmgren, 1860)
  pulchrator (Thomson, 1877)
  theridii (Howard, 1892)
  granulosa (Davis, 1898)
  decorata (Haupt, 1954)
  rhombifer (Haupt, 1954)

PIMPLINI
EPHIALTINI sensu Townes

ITOPLECTIS Foerster, 1869
alternans (Gravenhorst, 1829)
  spiracularis (Morley, 1908)
aterrima Jussila, 1965
  kolthoffi ?misident.
clavicornis (Thomson, 1889)
  curticauda misident.
insignis Perkins, 1957
maculator (Fabricius, 1775)
melanocephala (Gravenhorst, 1829)
  burtoni (Morley, 1946)

PIMPLA Fabricius, 1804
COCCYGOMIMUS Saussure *in* Grandidier, 1892
aethiops Curtis, 1828
arctica Zetterstedt, 1838
contemplator (Müller, 1776)
  rufistigma Morley, 1908
  rufitibia Morley, 1908
flavicoxis Thomson, 1877
  ?*aquilonia* Cresson, 1870
hypochondriaca (Retzius, 1783) **stat. nov.**
  inguinalis (Geoffroy *in* Fourcroy, 1785) **syn. nov.**
  instigator (Fabricius, 1793) homonym, **syn. nov.**
  processioniae Ratzeburg, 1849 **syn. nov.**
  intermedia Holmgren, 1860 **syn. nov.**

melanacrias Perkins, 1941
   geniculata Hensch, 1929 homonym
sodalis Ruthe, 1859
spuria Gravenhorst, 1829
turionellae (Linnaeus, 1758)
   examinator (Fabricius, 1804)
   opacellata Desvignes, 1868
wilchristi Fitton, Shaw & Gauld **sp. nov.**

APECHTHIS Foerster, 1869
   EPHIALTES Schrank, 1802 suppressed
compunctor (Linnaeus, 1758)
quadridentata (Thomson, 1877)
   resinator misident.
rufata (Gmelin *in* Linnaeus, 1790)

DELOMERISTINI
THERONIINI

THERONIA Holmgren, 1859
atalantae (Poda, 1761)

DELOMERISTA Foerster, 1869
mandibularis (Gravenhorst, 1829)
   albicinctus (Desvignes, 1862)
   desvignesii (Marshall, 1870)
novita (Cresson, 1870)
   laevis misident.
pfankuchi (Brauns, 1905)

PERITHOUS Holmgren, 1859
divinator (Rossius, 1790)
scurra (Panzer, 1804)
   mediator (Fabricius, 1804)
   ?*senator* (Haliday, 1838)

HYBOMISCHOS Baltazar, 1961
   HYMBOISCHOS: Fitton, 1978 misspelling
septemcinctorius (Thunberg, 1822)
   varius (Gravenhorst, 1829)

PSEUDORHYSSA Merrill, 1915
alpestris (Holmgren, 1859)

POEMENIINI
NEOXORIDINI

POEMENIA Holmgren, 1859
   CALLICLISIS Foerster, 1869
collaris Haupt, 1917
hectica (Gravenhorst, 1829)
notata Holmgren, 1859

DEUTEROXORIDES Viereck, 1914
elevator (Panzer, 1799)
   albitarsus (Gravenhorst, 1829)

PODOSCHISTUS Townes, 1957
scutellaris (Desvignes, 1856)

RHYSSINI

RHYSSA Gravenhorst, 1829
persuasoria (Linnaeus, 1758)

RHYSSELLA Rohwer, 1920
approximator (Fabricius, 1793)
curvipes (Gravenhorst, 1829)

DIACRITINI

DIACRITUS Foerster, 1869
PHIDIAS Vollenhoven, 1878
aciculatus (Vollenhoven, 1878)

Key to genera

Most of the European genera of Pimplinae occur in Britain. Confirmatory characters
which apply only to the British genera are included at appropriate points in the key (in
parentheses); so that any additional genera will be trapped rather than run spuriously
to a genus already recorded as British. In the construction of the key some allowance
was also made for non-British, European species, but not at the cost of overcomplicat-
ing it for British ones.

1   Mesoscutum almost entirely covered with sharp transverse ridges (Fig. 11). 2
—   Mesoscutum not covered with such ridges, if strongly sculptured usually punctate, rarely
     with a few transverse striae postero-medially. 4
2   Occipital carina mediodorsally complete (see Figs 40, 63, 64). Fore wing with cu-a at bifur-
     cation of M and Cu (Fig. 12). Female: last visible tergite of gaster not extended into a
     truncate horn (Fig. 14). . . . . **PSEUDORHYSSA** (p. 88)
—   Occipital carina mediodorsally effaced (see Fig. 39). Fore wing with cu-a joining Cu distal to
     bifurcation of M and Cu (Fig. 13). Female: last visible tergite of gaster extended into a
     truncate horn (Fig. 15). 3
3   Mid trochantellus without a lamella. Female: sternites 2–4 of gaster with median pair of
     tubercles (ovipositor guides) near their midlength (Fig. 17). Male: hind edge of tergites 3–6
     of gaster straight. . . . . **RHYSSA** (p. 91)
—   Mid trochantellus with a small lamella antero-ventrally at its apex (Fig. 16). Female:
     sternites 2–4 of gaster with median pair of tubercles near anterior margin (Fig. 18). Male:
     hind edge of tergites 3–6 of gaster concave (Fig. 19). (Female: tergites 3–5 of gaster with
     fine, transverse aciculate sculpture. Male: concave hind margin of tergites 3–6 of gaster
     simple, without a median elongate membranous area.). . . **RHYSSELLA** (p. 91)
4   Epicnemial carina entirely absent. Upper part of temple with or without scale-like
     ridges. 5
—   Epicnemial carina present ventrally and usually also laterally (as in Figs 1, 44, 45). Upper
     part of temple without scale-like ridges. 7
5   Mandible with two apical teeth (upper tooth much shorter than lower tooth).
     . . . . **POEMENIA** (p. 89)
—   Mandible with tip chisel-shaped, not divided into two teeth (Fig. 20). 6
6   Main sculpture of dorsal half of temple consisting of large, uneven, scale-like ridges (Fig. 21).
     (Tarsal claws with a subapical tooth.) . . . . . . **PODOSCHISTUS** (p. 90)
—   Main sculpture of dorsal half of temple consisting of minute, even, scale-like ridges
     (the individual ridges about the size of the smallest shown in Fig. 21).
     . . . . **DEUTEROXORIDES** (p. 90)

7	Segment 1 of gaster with sternite extending posteriorly at least 0·5 of distance between spiracle and posterior edge of tergite (Fig. 22); tergite more than 2·6 times as long as broad. Pronotum mediodorsally with a deep longitudinal groove. . **DIACRITUS** (p. 92)
—	Segment 1 of gaster with sternite extending posteriorly at very most 0·3 of distance between spiracle and posterior edge of tergite (Fig. 23); tergite less than 2·3 times as long as broad. Pronotum without a deep mediodorsal groove. 8
8	Fore wing with vein 3rs-m entirely absent (Fig. 24). 9
—	Fore wing with vein 3rs-m present, though occasionally unpigmented, thus delimiting a small cell (the areolet) (Fig. 25). 21
9	Clypeus and face confluent, forming a nearly flat surface (Fig. 26). (Surface of eyes bearing numerous long hairs which are more than three times the diameter of an eye facet in length.) . . . . **SCHIZOPYGA** (p. 62)
—	Clypeus and face separated by a distinct transverse impression (the clypeal suture) (Fig. 27). (Surface of eyes usually with only sparse, short hairs.) 10
10	Mesoscutum with a vertical carina on each side in front of the notaulus (Fig. 28). . . . . **ACRODACTYLA** (part) (p. 66)
—	Mesoscutum without distinct vertical carinae. 11
11	Tarsus with fifth segment conspicuously broadened and usually with pulvillus projecting beyond apices of claws (Fig. 30). Female with ovipositor in lateral view tapered from middle to the apex, at most only weakly curved (Fig. 34); lower valve with a weak to strong swelling at its midlength. 12
—	Tarsus with the fifth segment not greatly broadened; pulvillus not projecting beyond apices of claws (Fig. 29). Female with ovipositor in lateral view straight and tapered only at apex (usually with a nodus) (Figs 32, 33) OR tapered from middle and distinctly upcurved (Fig. 31); lower valve without a swelling at its midlength. 16
12	Notaulus absent. Gaster with tergites 2 to 4 evenly convex. . . **PIOGASTER** (p. 63)
—	Notaulus present (see Fig. 2). Gaster with tergites 2 to 4 with raised areas or rounded swellings, more or less defined by diagonal and transverse grooves (rather weak in one species of *Acrodactyla*) (Figs 37, 38).. 13
13	Hind wing with final abscissa of Cu present (Fig. 35) AND/OR mesoscutum with some areas entirely bare and the remainder with only sparse hairs AND/OR tergite 2 of gaster coriaceous or strongly punctured centrally. 14
—	Hind wing with final abscissa of Cu absent (Fig. 36) (very rarely with a weak impression in the wing membrane in the position the vein would occupy) AND mesoscutum with an even covering of hairs AND tergite 2 of gaster smooth and polished centrally. . . 15
14	Tergites of gaster (particularly 2–4) (Fig. 37) with the central area evenly convex, with almost no punctures, but ranging from coriaceous to smooth and polished; bounded by well defined posterior transverse and anterolateral diagonal grooves. **ZATYPOTA** (p. 68)
—	Tergites of gaster (particularly 2–4) (Fig. 38) with a pair of poorly to strongly developed rounded swellings, usually covered with fine to strong punctures, especially medially; bounded by weak depressions. (Surface of eyes almost bare. Tergites of gaster with relatively fewer and much finer punctures on the sublateral swellings than medially. Female: ovipositor-hind tibia index at least 0·7.). . . . . . **POLYSPHINCTA** (p. 64)
15	Occipital carina interrupted mediodorsally (Fig. 39). Tergite 1 of gaster less than 1·5 times as long as broad. . . . . **SINARACHNA** (p. 68)
—	Occipital carina complete mediodorsally (Fig. 40). Tergite 1 of gaster more than 1·7 times as long as broad. . . . . **ACRODACTYLA** (part) (p. 66)
16	Female.. 17
—	Male. 19
17	Ovipositor in lateral view distinctly upcurved and tapered from middle to apex (Fig. 31). . . . . **CLISTOPYGA** (female) (p. 60)
—	Ovipositor in lateral view straight and tapered only at apex (Figs 32, 33). 18
18	Proximal tooth of lower valve of ovipositor large, forming a long antero-dorsal spine (Fig. 32) (the spine is sometimes broken off and the tip of the ovipositor should be examined from dorsal and lateral views).. . . . . **ZAGLYPTUS** (female) (p. 58)
—	Proximal tooth of lower valve of ovipositor not enlarged, not forming an antero-dorsal spine (Fig. 33).. . . . . **TROMATOBIA** (female, part) (p. 56)
19	Head behind malar space with a groove bounded dorso-laterally by a ridge (Fig. 41). Antennal segments lacking tyloids AND hind wing with abscissa of Cu between M + Cu and cu-a longer than cu-a (as in Fig. 35). . . . . **CLISTOPYGA** (male) (p. 60)

— Head behind malar space without a groove bounded by a ridge. Some of antennal segments 6 to 10 bearing tyloids (Figs 42, 43) AND/OR hind wing with abscissa of Cu between M + Cu and cu-a at most as long as cu-a. 20

20 Antenna with tyloids on outer side of segments 6 and 7 (Fig. 42) or 8 to 10 (Fig. 43). . . .
. . . . **ZAGLYPTUS** (male) (p. 58)

— Antenna without tyloids. . . . . **TROMATOBIA** (male, part) (p. 56)

21 Mesopleural furrow not angled opposite episternal scrobe (Fig. 44). (Notauli absent or present only anteriorly and relatively weak. Male subgenital plate long and posteriorly more-or-less pointed (Fig. 47). Hind wing with abscissa of Cu between M + Cu and cu-a at most 0·6 as long as cu-a. Ovipositor-hind tibia index not more than about 1·0) . . . 22

— Mesopleural furrow angled opposite episternal scrobe (Fig. 45). 26

22 Female. 23

— Male. 24

23 Ovipositor tip abruptly down-curved (Fig. 46). Face black, usually with orbits narrowly yellow. . . . . **APECHTHIS** (female) (p. 83)

— Ovipositor straight. Face entirely black. 25

24 Gaster tergite 5 with laterotergite narrow, at least 4 times as long as broad (Fig. 48). Face extensively pale-marked. . . . . **APECHTHIS** (male) (p. 83)

— Gaster tergite 5 with laterotergite wide, at most 3 times as long as broad (Fig. 49). Face entirely black. 25

25 Hind tarsus with segment 2 distinctly shorter than segment 5 (Fig. 50) and margin of eye strongly indented just above level of antennal socket (Fig. 51). Antenna often subclavate (Figs 53, 54). Female: fore tarsal claws often with a tooth-like basal lobe (Fig. 56). . . . . **ITOPLECTIS** (p. 70)

— Hind tarsus with segment 2 just shorter than, equal to, or longer than segment 5 and margin of eye usually only weakly concave just above level of antennal socket (Fig. 52). Antenna not subclavate (often attenuate apically (Fig. 55)). Female: fore tarsal claws without a tooth-like basal lobe (Fig. 57). (Antenna slender: flagellum segment 2 at least 3·0 times as long as broad. Female: ovipositor projecting conspicuously beyond end of gaster.) . . . . **PIMPLA** (p. 74)

26 Tergites 2–4 of gaster polished, unsculptured. All tarsal claws very large and with a prominent, flattened, spatulate-tipped bristle (Fig. 58). . . . . . **THERONIA** (p. 84)

— Tergites 2–4 matt to polished, punctate or otherwise sculptured. Tarsal claws not as enlarged (Figs 59–61) and not bearing a bristle so conspicuously modified (at most as in Fig. 59). 27

27 Female. 28

— Male. 47

28 Fore tarsal claw without a tooth-like basal lobe (see Figs 57–59). (Clypeus without a median apical tubercle.) 29

— Fore tarsal claw with a tooth-like basal lobe (Figs 60, 61). 32

29 Gastral tergites irregularly granulate, sometimes punctate-reticulate anteriorly, submatt. Ovipositor-hind tibia index at most 2·1. Propodeum with area superomedia distinct, defined by transverse and longitudinal carinae (Fig. 62). . . . .
. . . . **DELOMERISTA** (female) (p. 85)

— Gastral tergites punctate, usually with distinct spaces between punctures and usually polished. Ovipositor-hind tibia index more than 2·4. Area superomedia not discernible on propodeum, at most with an indication of longitudinal carinae. 30

30 Occipital carina mediodorsally dipped (Fig. 63). Orbits entirely black.
. . . . **EXERISTES** (female) (p. 44)

— Occipital carina mediodorsally evenly convex (Fig. 64). Orbits whitish or yellow striped.
. . . . 31

31 Tergite 1 of gaster with antero-lateral corner produced to form a tooth (Fig. 65). Tip of ovipositor sinuate (Fig. 67). . . . . . . **HYBOMISCHOS** (female) (p. 87)

— Tergite 1 of gaster with antero-lateral corner simple (Fig. 66). Tip of ovipositor not sinuate.
. . . . **PERITHOUS** (female) (p. 86)

32 Occipital carina mediodorsally evenly convex (see Fig. 64) or horizontal, always complete. Edge of clypeus truncate or slightly concave (see Fig. 41). (Small to moderate sized species, fore wing length less than 9·0 mm.) 33

— Occipital carina mediodorsally dipped (Fig. 63), sometimes obsolescent (see Fig. 39).

Clypeus usually with apical margin impressed, making it conspicuously bilobed (see Fig. 83). (Small to large sized species.). 34

33 Ovipositor in lateral view tapered from middle to apex (see Fig. 34). Tarsus with fifth segment conspicuously broadened (Fig. 30). Hind wing with abscissa of Cu between M + Cu and cu-a at least 1·3 as long as cu-a.. . . . . **DREISBACHIA** (female) (p. 62)
— Ovipositor in lateral view tapered only at apex (Fig. 33). Tarsus with fifth segment not greatly broadened (Fig. 29). Hind wing with abscissa of Cu between M + Cu and cu-a at most 1·2 as long as cu-a.. . . . . . . **TROMATOBIA** (female, part) (p. 56)

34 Ovipositor–hind tibia index less than 2·9. 35
— Ovipositor–hind tibia index more than 3·0 (specimens with the index intermediate will key either way). 40

35 Ovipositor tip with lower valves extended to partially enclose upper valve and clearly visible in dorsal view (Fig. 68). . . . . **FREDEGUNDA** gen. nov. (female) (p. 45)
— Ovipositor tip with lower valves not at all enclosing the upper valve (Figs 69, 70). . . 36

36 Ovipositor tip with teeth, excluding the most proximal one, more or less vertical (Fig. 69). Mid tarsus with segment 5 at least 0·9 times as long as segment 1. Hind tarsal claws usually with apex bevelled (Fig. 60). . . . . . . **ENDROMOPODA** (female) (p. 45)
— Ovipositor tip with central and proximal teeth oblique (Fig. 70). Mid tarsus with segment 5 less than 0·9 times as long as segment 1. Hind tarsal claws with apex simply pointed (Fig. 61).. 37

37 Face black with a yellow triangular mark below each antennal socket. (Fore wing usually with 2m-cu meeting M almost adjacent to 3rs-m, and usually with 2rs-m and 3rs-m joining before meeting Rs (the combined vein being longer than wide). Hind wing usually with abscissa of Cu between M + Cu and cu-a more than 1·1 as long as cu-a.) . . . . **ACROPIMPLA** (female) (p. 55)
— Face entirely black beneath the antennal sockets with, at most, only the clypeus yellowish. 38

38 Hind wing with abscissa of Cu between M + Cu and cu-a longer than cu-a (Fig. 72). . . . . **SCAMBUS** (female, part) (p. 48) (A few specimens of **GREGOPIMPLA** will also run here;they are catered for in the key to *Scambus* species.)
— Hind wing with abscissa of Cu between M + Cu and cu-a shorter than cu-a (Fig. 71).. 39

39 Propodeum in profile weakly convex (Fig. 73). Pterostigma black or dark brown (usually pale at extreme proximal corner). Tergite 2 of gaster with oblique depressions cutting off anterolateral corners, and with lateromedian swellings that are less densely punctate than is the central area.. . . . . **ISEROPUS** (female) (p. 56)
— Propodeum in profile more strongly convex (Fig. 74). Pterostigma yellowish. Tergite 2 of gaster without distinct oblique depressions anterolaterally, generally rather evenly convex and centrally broadly uniformly punctate. . **GREGOPIMPLA** (female, part) (p. 56)

40 Ovipositor–hind tibia index less than 3·7 AND hind wing with abscissa of Cu between M + Cu and cu-a at least as long as cu-a (Fig. 72). 41
— Ovipositor–hind tibia index more than 3·9 AND/OR hind wing with abscissa of Cu between M + Cu and cu-a shorter than cu-a (Fig. 71). 42

41 Ovipositor tip with lower valves extended to partially enclose the upper valve and clearly visible in dorsal view (Figs 75, 76 and see Fig. 68). . . . . **DOLICHOMITUS** (female, part) (p. 36)
— Ovipositor tip with lower valves not at all enclosing the upper valve (Figs 78, 79 and see Fig. 70). . . . . **SCAMBUS** (female, part) (p. 48)

42 Mandible long, with lower tooth about twice as long as upper one (Fig. 80). Face broad, more than 1·7 times as wide as high (from antennal socket to clypeal suture) (Fig. 83). (Ovipositor tip with lower valves extended to partially enclose the upper valve and clearly visible in dorsal view (see Figs 68, 75, 76). Eyes with inner orbits diverging from just below antennal sockets (Fig. 83).). . . . . **EPHIALTES** (female) (p. 35)
— Mandible shorter, with lower tooth at most 1·5 times length of upper one, generally subequal (Figs 81, 82). Face less broad, at most 1·6 times as wide as high. 43

43 Submetapleural carina represented only anteriorly (Fig. 86) AND scutellum black AND with anterior punctured area of tergite 2 about 3 times length of smooth posterior margin (Fig. 84) AND tergite 3 1·8–2·0 times as long as broad. (Fore wing with cross vein cu-a meeting M + Cu before bifurcation of M and Cu. Hind tarsus with segment 3 at least 1·5 times as long as segment 5.). . . . . **TOWNESIA** (female) (p. 40)

— Submetapleural carina reaching posteriorly to metacoxa (Fig. 87) OR, if not (Fig. 88), either with scutellum yellow marked or with anterior punctured area of tergite 2 more than 3·5 times length of smooth posterior margin or tergite 3 less than 1·8 times as long as broad.
. . . . 44

44 Ovipositor tip (Figs 75, 76) with lower valves extended to partially enclose the upper valve and clearly visible in dorsal view; upper valve never with a distinct row of dorsolateral minute teeth. . . . . **DOLICHOMITUS** (female, part) (p. 36)

— Ovipositor tip (see Figs 77–79) with lower valve not at all enclosing the upper valve; upper valve often with a dorsolateral row of minute teeth (which can be difficult to see) (Fig. 77).
. . . . 45

45 Mandible conspicuously puncto-striate on proximal half. Tergite 2 with strongly impressed grooves cutting off anterior corners AND hind coxa red. Sternite 1 of gaster more than 1·5 times as long medially as posteriorly broad. . **PARAPERITHOUS** (female) (p. 40)

— Mandible without obvious striae, although it may be otherwise strongly sculptured or punctate. Tergite 2 with at most relatively weak oblique grooves, except in one species with the hind coxa blackish (hind coxa usually red). Sternite 1 of gaster less than 1·2 times as long as broad. 46

46 Tergite 2 of gaster with a broad impunctate posterior rim which is at least as long as 6 times diameter of central punctures AND hind tarsus segment 5 subequal to or shorter than segment 3 AND propodeum with lateromedian longitudinal carinae indistinct or absent.
. . . . **LIOTRYPHON** (female) (p. 41)

— Tergite 2 of gaster punctate almost to hind margin, the impunctate posterior rim as long as 3 times diameter of central punctures, AND/OR hind tarsus segment 5 more than 1·5 times as long as segment 3 AND/OR propodeum with lateromedian longitudinal carina distinct, at least anteriorly. (Lateromedian longitudinal carinae present on only the anterior part of the dorsal surface of the propodeum.) . . **AFREPHIALTES** (female) (p. 44)

47 Tergite 1 of gaster much narrower posteriorly than at the level of the spiracles (Fig. 85).
. . . . **FREDEGUNDA** gen. nov. (male) (p. 45)

— Tergite 1 of gaster not markedly narrowed posteriorly (e.g. Fig. 84). 48

48 Face (and usually also the clypeus) at least partly white, cream or yellow. (Clypeus without a median apical tubercle.) 49

— Face entirely black or blackish (clypeus may not be black). 55

49 Occipital carina mediodorsally evenly convex (Fig. 64) or horizontal, always complete.
. . . . 50

— Occipital carina mediodorsally dipped (Fig. 63), someimes obsolescent (see Fig. 39). (Subgenital plate short, its posterior edge concave.) 54

50 Clypeus conspicuously rounded and convex (see Fig. 27), its margin not impressed medially. Subgenital plate short, its posterior edge concave or convex (Figs 89, 90). . . . 51

— Clypeus flat or with a transverse raised area and often with the margin impressed medially (see Fig. 83). Subgenital plate long, its posterior edge strongly convex (Fig. 91).. . 52

51 Posterior edge of subgenital plate convex (Fig. 89). Tarsus with fifth segment conspicuously broadened (Fig. 30). Hind wing with abscissa of Cu between M + Cu and cu-a at least 1·3 as long as cu-a.. . . . . **DREISBACHIA** (male) (p. 62)

— Posterior edge of subgenital plate concave (Fig. 90). Tarsus with fifth segment not greatly broadened (Fig. 29). Hind wing with abscissa of Cu between M + Cu and cu-a at most 1·2 as long as cu-a.. . . . . **TROMATOBIA** (male, part) (p. 56)

52 Gastral tergites irregularly granulate, sometimes punctate-reticulate anteriorly, submatt. Propodeum with area superomedia distinct, defined by transverse and longitudinal carinae (Fig. 62). . . . . **DELOMERISTA** (male) (p. 85)

— Gastral tergites punctate, at least tergites 3–5 with spaces between punctures polished and shining. Area superomedia not discernible on propodeum, at most with an indication of longitudinal carinae. 53

53 Tergite 1 of gaster with antero-lateral corner produced to form a tooth (Fig. 65). . . . . **HYBOMISCHOS** (male) (p. 87)

— Tergite 1 of gaster with antero-lateral corner simple (Fig. 66). . . . . **PERITHOUS** (male) (p. 86)

54 Hind tibia whitish, blackish distally and with a sub-proximal blackish band; hind tarsus with each segment whitish proximally and blackish distally. Hind wing usually with abscissa of Cu between M + Cu and cu-a less than 0·9 as long as cu-a. Fore wing usually with 2m-cu

meeting M about two-thirds between 2rs-m and 3rs-m, and usually with 2rs-m and 3rs-m not joining before meeting Rs (although they are adjacent). . . . .
. . . . **ISEROPUS** (male) (p. 56)

— Hind tibia whitish, blackish distally, without a sub-proximal blackish band; hind tarsus (not each segment individually) whitish proximally and blackish distally. Hind wing usually with abscissa of Cu between M + Cu and cu-a more than 1·1 as long as cu-a. Fore wing usually with 2m-cu meeting M almost adjacent to 3rs-m, and usually with 2rs-m and 3rs-m joining before meeting Rs (the combined vein being longer than wide).
. . . . **ACROPIMPLA** (male) (p. 55)

55 Fore wing with very long hairs on the veins forming the leading edge proximal to the pterostigma (the hairs sharply curved at their tip and conspicuously longer than the width of the combined veins (C + Sc + R + Rs)). 56

— Fore wing with the hairs on the veins forming the leading edge not abnormally long (more or less straight and as long as about half the width of the combined veins (C + Sc + R + Rs)).
. . . . 57

56 Mandibles long and tapered, with lower tooth about twice as long as upper one (Fig. 80). Clypeus whitish. Fore wing with very long hairs restricted to the leading edge.
. . . . **EPHIALTES** (male) (p. 35)

— Mandible shorter and less tapered, with teeth subequal. Clypeus blackish or brownish. Fore wing with very long hairs on vein M + Cu as well as on the leading edge.
. . . . **TOWNESIA** (male) (p. 40)

57 Occipital carina mediodorsally evenly convex and complete (Fig. 64). (Clypeus rounded and convex (see Fig. 27).). . . . . **TROMATOBIA** (male, part) (p. 56)

— Occipital carina mediodorsally dipped (Fig. 63), sometimes obsolescent (see Fig. 39). 58

58 Mandibles conspicuously puncto-striate on proximal half. (Tergite 2 with strongly impressed grooves cutting off anterior corners. Hind wing with abscissa of Cu between M + Cu and cu-a much shorter than cu-a.). . **PARAPERITHOUS** (male) (p. 40)

— Mandible without obvious striae, although it may be otherwise strongly sculptured or punctate. 59

59 **GREGOPIMPLA** (male) (p. 56), **ENDROMOPODA** (male) (p. 45), **SCAMBUS** (male) (p. 48), **DOLICHOMITUS** (male) (p. 36), **LIOTRYPHON** (male) (p. 41), **AFREPHIALTES** (male) (p. 44) and **EXERISTES** (male) (p. 44) run here. Although there are aggregate differences between males of some of these genera, and several individual species can be characterised, it has not proved possible to devise a practical dichotomous key to separate them at generic level. Some species can best be identified by association with females; either through collecting and/or rearing or through the characters of individual species. Also the states of the three characters given in the table below can be used as a guide to possible generic placement. The characters are:

1. (Range of) length to breadth of gaster tergite 2.
2. Condition of grooves cutting off anterolateral corners of gaster tergite 2.
3. Length of abscissa of Cu between M + Cu and cu-a compared to length of cu-a in hind wing.

| *Genus* | *Characters* | | |
	1	2	3
ENDROMOPODA	0·9–1·7	absent–faint	just shorter–longer
SCAMBUS	0·7–1·4	absent–faint	longer
GREGOPIMPLA	0·9–1·1	absent–faint	shorter or as long
LIOTRYPHON	0·9–2·9	absent–weak	shorter or as long
DOLICHOMITUS	1·3–2·3	weak–strong	shorter–longer
EXERISTES	1·0–1·2	very weak	just shorter–longer
AFREPHIALTES	ca. 0·8	very weak	shorter

Notes on particular genera:
ENDROMOPODA. A number of species have the underside of the fore femur modified (see Figs 111, 112).
SCAMBUS. A number of species have the underside of the fore femur modified (as in *Endromopoda*, see Figs 111, 112).
DOLICHOMITUS. In *mesocentrus* the mid coxa is curiously modified (Fig. 97).

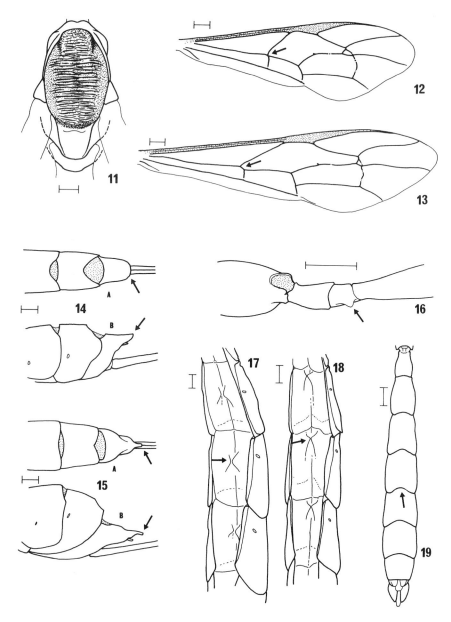

Figs 11–19. 11, mesoscutum, dorsal view, *Pseudorhyssa alpestris*. 2, right fore wing, *Pseudorhyssa alpestris*. 13, right fore wing, *Rhyssella approximator*. 14. apex of gaster, (A) dorsal view, (B) lateral view, *Pseudorhyssa alpestris* female. 15, apex of gaster, (A) dorsal view, (B) lateral view, *Rhyssella approximator* female. 16, trochanter and trochantellus of right mid leg, front view, *Rhyssella approximator*. 17, segments 2–4 of gaster, latero-ventral view, *Rhyssa persuasoria* female. 18, segments 2–4 of gaster, latero-ventral view, *Rhyssella approximator* female. 19, gaster, dorsal view, *Rhyssella approximator* male. All scale lines represent 0·5 mm.

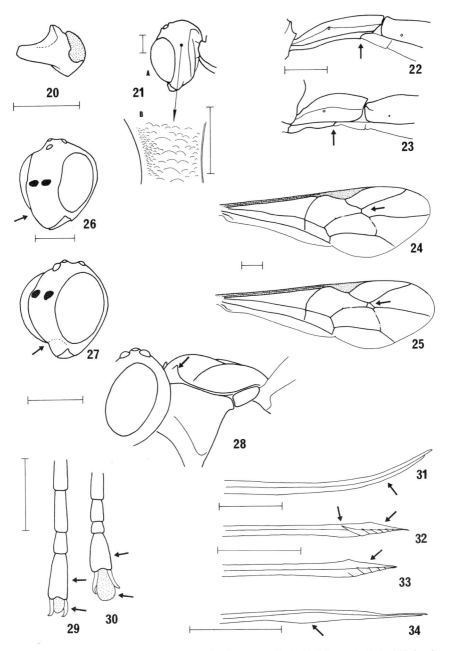

Figs 20–34. 20, left mandible, front view, *Podoschistus scutellaris*. 21, left temple, (A) whole head to show scale, (B) area of temple between eye and occipital carina, *Podoschistus scutellaris*. 22, segment 1 of gaster, lateral view, *Diacritus aciculatus*. 23, segment 1 of gaster, lateral view, *Perithous divinator*. 24, right fore wing, *Polysphincta boops*. 25, right fore wing, *Tromatobia oculatoria*. 26, head, antero-lateral view, *Schizopyga frigida*. 27, head, antero-lateral view, *Acrodactyla quadrisculpta*. 28, mesoscutum, antero-lateral view, *Acrodactyla quadrisculpta*. 29, apex of hind tarsus, dorsal view, *Tromatobia oculatoria*. 30, apex of hind tarsus, dorsal view,· *Dreisbachia pictifrons*. 31, apex of ovipositor, lateral view, *Clistopyga incitator*. 32, apex of ovipositor, lateral view, *Zaglyptus varipes*. 33, apex of ovipositor, lateral view, *Tromatobia ovivora*. 34, apex of ovipositor, lateral view, *Sinarachna pallipes*. All scale lines represent 0·5 mm.

29

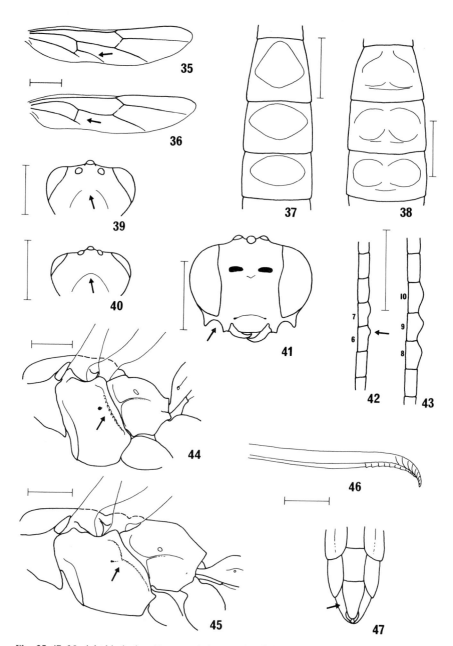

Figs 35–47. 35, right hind wing, *Zatypota bohemani*. 36, right hind wing, *Sinarachna pallipes*. 37, tergites 2–4 of gaster, dorsal view, *Zatypota percontatoria*. 38, tergites 2–4 of gaster, dorsal view, *Polysphincta tuberosa*. 39, upper part of head, posterior view, *Sinarachna pallipes*. 40, upper part of head, posterior view, *Acrodactyla degener*. 41, head, front view, *Clistopyga incitator* male. 42, segments 5–9 of right antenna (3–7 of flagellum), dorsal view, *Zaglyptus multicolor* male. 43, segments 7–11 of right antenna, dorsal view, *Zaglyptus varipes* male. 44, meso- and metathorax and propodeum, lateral view, *Pimpla wilchristi*. 45, meso- and metathorax and propodeum, lateral view, *Endromopoda detrita*. 46, apex of ovipositor, lateral view, *Apechthis compunctor*. 47, apex of gaster, ventral view, *Pimpla wilchristi*, male. All scale lines represent 0·5 mm.

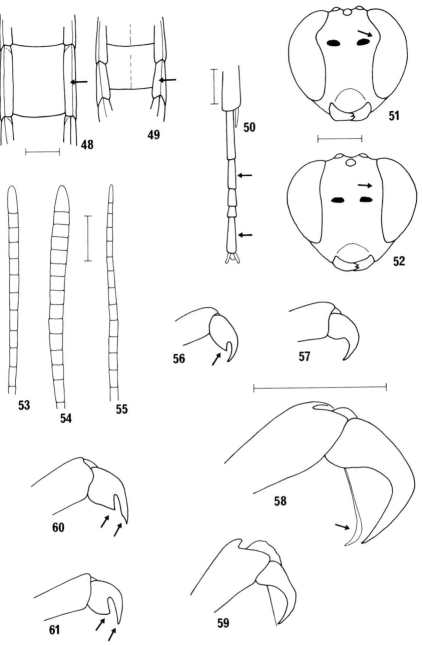

Figs 48–61. 48, segment 5 of gaster, ventral view, *Apechthis rufata* male. 49, segment 5 of gaster, ventral view, *Pimpla flavicoxis* male. 50, hind tarsus, dorsal view, *Itoplectis maculator*. 51, head, front view, *Itoplectis maculator*. 52, head, front view, *Pimpla wilchristi*. 53, apex of antenna, *Itoplectis maculator* female. 54, apex of antenna, *Itoplectis clavicornis* female. 55, apex of antenna, *Pimpla flavicoxis* female. 56, fore tarsal claw, lateral view, *Itoplectis maculator* female. 57, fore tarsal claw, lateral view, *Pimpla flavicoxis* female. 58, tarsal claw, lateral view, *Theronia atalantae* male. 59, tarsal claw, lateral view, *Fredegunda diluta* male (only the longest and strongest bristle shown). 60, tarsal claw, lateral view, *Endromopoda arundinator* female. 61, tarsal claw, lateral view, *Scambus nigricans* female. All scale lines represent 0·5 mm.

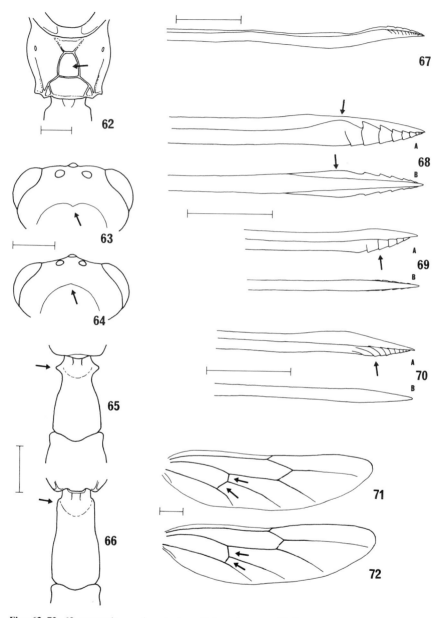

Figs 62–72. 62, propodeum, dorsal view, *Delomerista mandibularis*. 63, upper part of head, posterior view, *Exeristes ruficollis*. 64, upper part of head, posterior view, *Perithous scurra*. 65, tergite 1 of gaster, dorsal view, *Hybomischos septemcinctorius*. 66, tergite 1 of gaster, dorsal view, *Perithous scurra*. 67, apex of ovipositor, lateral view, *Hybomischos septemcinctorius*. 68, apex of ovipositor, (A) lateral view, (B) dorsal view, *Fredegunda diluta*. 69, apex of ovipositor, (A) lateral view, (B) dorsal view, *Endromopoda arundinator*. 70, apex of ovipositor, (A) lateral view, (B) dorsal view, *Scambus brevicornis*. 71, right hind wing, *Iseropus stercorator*. 72, right hind wing, *Scambus brevicornis*. All scale lines represent 0·5 mm.

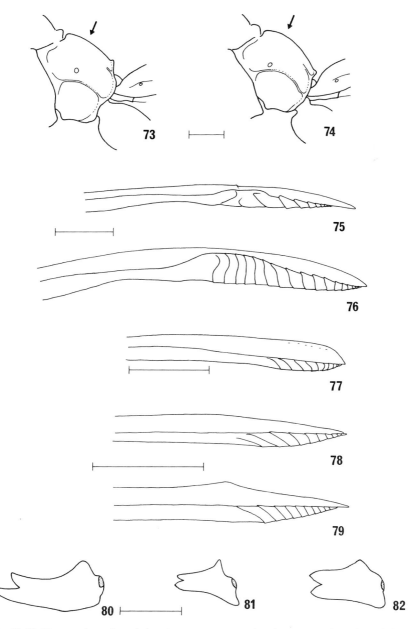

Figs 73–82. 73, propodeum, lateral view, *Iseropus stercorator* female. 74, propodeum, lateral view, *Gregopimpla inquisitor* female. 75, apex of ovipositor, lateral view, *Dolichomitus imperator*. 76, apex of ovipositor, lateral view, *Dolichomitus tuberculatus*. 77, apex of ovipositor, lateral view, *Liotryphon caudatus*. 78, apex of ovipositor, lateral view, *Scambus sagax*. 79, apex of ovipositor, lateral view, *Scambus buolianae*. 80, left mandible, front view, *Ephialtes manifestator*. 81, left mandible, front view, *Dolichomitus agnoscendus*. 82, left mandible, front view, *Dolichomitus terebrans*. All scale lines represent 0·5 mm.

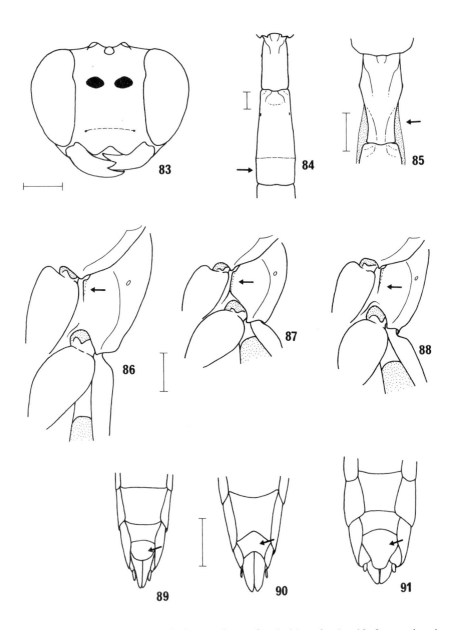

Figs 83–91. 83, head, front view, *Ephialtes manifestator* female. 84, tergites 1 and 2 of gaster, dorsal view, *Townesia tenuiventris*. 85, tergite 1 of gaster, dorsal view, *Fredegunda diluta* male. 86, metapleuron, ventro-lateral view, *Townesia tenuiventris* female. 87, metapleuron, ventro-lateral view, *Liotryphon caudatus* female. 88, metapleuron, ventro-lateral view, *Liotryphon ruficollis* female. 89, apex of gaster, ventral view, *Dreisbachia pictifrons* male. 90, apex of gaster, ventral view, *Tromatobia oculatoria* male. 91, apex of gaster, ventral view, *Delomerista mandibularis* male. All scale lines represent 0·5 mm.

Tribe EPHIALTINI

Worldwide this tribe comprises about 30 genera, of which 18 occur in Europe. Sixteen of these, including *Fredegunda* described below, are known from Britain. Of the two remaining European genera, one, *Pseudopimpla*, is Mediterranean in distribution and parasitises hartigiine sawflies (Bruzzese, 1982). The other, *Alophosternum*, is more widely distributed and in Europe is known to attack heterarthrine sawflies (Zwakhals, 1987), while in North America it has a wider range of hosts including leaf-mining Lepidoptera (Carlson, 1979). Townes (1969) divided the Ephialtini into four groups of genera and all of the British genera belong to the *Ephialtes*-group (*Pimpla*-group sensu Townes). The tribe includes some of the morphologically least specialised of all Pimplinae and, as currently constituted, is undoubtedly paraphyletic. The genera *Clistopyga, Tromatobia* and *Zaglyptus* form a link with the Polysphinctini but otherwise relationships between tribes are not clear.

Except that *Clistopyga, Tromatobia* and *Zaglyptus* are associated with spiders' nests, the genera of this tribe develop as ectoparasitoids of holometabolous insect larvae, prepupae or pupae that are more or less fully concealed: in wood or stems, under bark, in galled, mined or spun vegetation, or in cocoons. These hosts are, as far as is known, stung and paralysed or killed prior to oviposition.

Genus EPHIALTES Gravenhorst

A small Holarctic genus represented in Europe by three or four species. Only one, *E. manifestator*, has been recorded as British. In British material the relative length of the ovipositor is very variable and to some extent this correlates with variation in other characters such as the shape or the first segment of the gaster. However, it has not proved possible to recognise distinct segregates. In this genus females are difficult to determine with certainty and unfortunately few males are known from Britain. Perhaps more than one of the European species occurs in Britain or, alternatively, *manifestator* as currently recognised may be a complex of species.

In the past the genus *Ephialtes* encompassed most of the long-ovipositored members of the tribe Ephialtini. *Ephialtes* in its current sense is characterised by the broad face, long tapered mandible with a very elongate lower tooth, and by the long hairs on the proximal, leading edge of the fore wing in males. Some European *Dolichomitus* show evidence of close relationship to *Ephialtes* (see page 36).

The generic name *Ephialtes* was first applied by Schrank to the genus now known as *Apechthis*. That application is still considered valid by some workers (e.g. Townes, 1969), who also apply the name *Pimpla* to *Ephialtes* as used here (see Fitton & Gauld, 1976). However, in Opinion 159 (1945), the International Commission on Zoological Nomenclature attributed authorship of *Ephialtes* to Gravenhorst (1829) and fixed the type species as *Ichneumon manifestator* Linnaeus.

— Moderately large to large species (although some males quite small), fore wing length 4·8–12·3 mm (in one exceptionally small male 2·8 mm). Ovipositor rather variable in length, ovipositor–hind tibia index 5·3–9·1. Mandible as in Fig. 80. Face broad, as in Fig. 83. Body black, hind corner of pronotum and tegula yellowish, legs mainly red.
. . . . **manifestator** (Linnaeus)

E. manifestator. Uncommon; southern England and Wales: as far west as Devon and as far north as Gwynedd, Staffordshire and Suffolk; Ireland: Kerry. Flight period: v–ix. In the past, many species of large Ephialtini with long ovipositors have been confused under this name. The many published records (see Aubert, 1969) must therefore be treated with even more than the usual suspicion. It is clear that *E. manifestator* attacks hosts that are rather deeply concealed in long-dead and sometimes rotten wood, and it has on several occasions been observed probing old emergence holes of wood-boring beetles in dead trees and fence posts. These observations suggest that aculeate Hymenoptera may be among the regular hosts, as is borne out by the rearing of a single very small male from *Trypoxylon* sp. (Sphecidae) in a trap nest, and by the recorded biology of North American congeners (Townes & Townes, 1960; Carlson, 1979). We have also seen a specimen labelled as reared from the cerambycid beetle *Callidium violaceum* (Linnaeus).

Genus **DOLICHOMITUS** Smith

Dolichomitus is a large Holarctic and South American genus with about 20 European species, of which 9 are known to occur in Britain. The limits of the genus are difficult to define. In keys to genera (e.g. Townes & Townes, 1960; Constantineanu & Pisica, 1977) a great deal of emphasis has been placed on the presence or absence of diagonal furrows on tergite 2 of the gaster. However, these grooves are rather weak in some species (e.g. *D. terebrans*, which otherwise is typical of the genus) and, conversely, are quite well developed in some members of related genera (e.g. *Liotryphon strobilellae*). The most characteristic feature of *Dolichomitus* seems to be the enlarged lobes of the lower ovipositor valves. This appears to be an adaptation enabling them to drill through bark to reach insects that have mandibulate adults (e.g. beetles) and which therefore pupate in bark or wood without first preparing a weak emergence cap.

The genus most likely to be confused with *Dolichomitus* is *Liotryphon*, owing mainly to the attempt to define polythetic taxa (see Gauld & Mound, 1982) in simple terms. The characters of *Dolichomitus* and *Liotryphon* are as follows:

Dolichomitus	*Liotryphon*
Ovipositor with lobes of lower valve:	
extended to partially	not at all enclosing upper
enclose upper valve;	valve; upper valve usually
upper valve without teeth	with a row of minute teeth
Gaster tergite 2 with diagonal furrows:	
usually deeply impressed	absent or weak
Propodeum with:	
usually long lateromedian	very short lateromedian
longitudinal carinae or a	longitudinal carinae or no
deep furrow anteriorly	furrow or carinae
Gaster tergite 1:	
usually very long,	usually quite short,
posteriorly often	posteriorly more evenly
irregularly sculptured,	punctate,
often concave	convex
Pronotum with postero-dorsal corner usually:	
black	with a yellow stripe

Individual species of either genus may disagree with one or two of these character states. The placement of at least one species is problematic: *D. agnoscendus* is sometimes regarded as a *Liotryphon* (e.g. by Oehlke, 1967), but more often as a *Dolichomitus*. Because of the long lower mandibular tooth Perkins (1943) associated *agnoscendus* (together with *subglabratus* Perkins and *lateralis* Wollaston) with *Ephialtes manifestator*. We regard all three former species as *Dolichomitus*, but the relationship of *Ephialtes* to all the Holarctic species currently placed in *Dolichomitus* is in need of further study. The difference between these two groups seems to be less than that between the Holarctic and Neotropical species currently classified as *Dolichomitus*. Constantineanu & Pisica (1977) recognised some of the species-groups of *Dolichomitus* used by Townes & Townes (1960) as distinct genera. We have opted to retain the classification of the European species given by Aubert (1969) and think that further splitting or other reorganisation of the classification is premature. Much more needs to be learned about the biology of these striking, large insects. Except as noted, all species are predominantly black with reddish legs.

Key to species (females only)

1   Lobe of lower valve of ovipositor with proximal ridges (2 or 3) widely separated (Fig. 92). 2
—   Lobe of lower valve of ovipositor with proximal ridges (4 or more) close together (Fig. 93). 3

2   Hind leg with tibia about as long as segments 1–4 of tarsus. Tergite 2 of gaster irregularly trans-striate, the setiferous punctures only weakly distinguished from the striae. Pterostigma dark brown with margins blackish. (Large to very large species, fore wing length 13·0–22·0 mm. Ovipositor–hind tibia index 4·8–5·7.) . . . . **imperator** (Kriechbaumer)
—   Hind leg with tibia subequal in length to the entire tarsus. Tergite 2 of gaster with punctures distinct from any weak transverse striae, fairly regularly punctate in at least the anterior half. Pterostigma pale brown with anterior margin dark. (Large species, fore wing length 10·1–16·4 mm. Ovipositor–hind tibia index 3·7–4·2.) . . . . . **pterelas** (Say)

3   Epicnemial carina extending laterally only just above level of lower corner of pronotum and not turning forwards to approach front margin of mesopleuron. (Moderately large species, although some males quite small, fore wing length 5·1–14·2 mm. Ovipositor–hind tibia index 3·5–4·2. Mandible with lower tooth not obviously longer than upper. Pterostigma yellow-brown. Lower part of mesopleuron sometimes reddish.). . . . . **populneus** (Ratzeburg)
—   Epicnemial carina extending well above level of lower corner of pronotum (the upper part occasionally less distinct but definitely turning forward to approach front margin of mesopleuron) (see Fig. 45).. 4

4   Propodeum with lateromedian longitudinal carinae strongly divergent, so that distance between them posteriorly is more than 3 times that anteriorly (Fig. 94). (Pterostigma blackish, sometimes a little paler centrally. Large species, fore wing length 8·9–15·0 mm. Ovipositor–hind tibia index 4·0–4·3.) . . . . **diversicostae** (Perkins)
—   Propodeum with lateromedian longitudinal carinae, which may be obsolete, more weakly diverging, the distance between them posteriorly at most about twice that anteriorly. 5

5   Pterostigma short and broad (Fig. 95), so that 2r-rs is less than 3·3 times as long as pterostigma is broad (if more than 2·9 then with lower tooth of mandible obviously longer than upper, Fig. 99). 6
—   Pterostigma narrower, so that 2r-rs is 3·5 or more times as long as pterostigma is broad (Fig. 96). (Mandible always with lower tooth about same length as upper.) 7

6   Mandible with lower tooth about as long as upper (Fig. 98). Tergite 2 of gaster with diagonal grooves quite weak, subtending an angle of about 45 degrees to anterior margin. (Moderately large species, fore wing length 6·0–10·1 mm. Ovipositor–hind tibia index 3·7–4·4. Pterostigma brownish with margins dark.) . . . . **terebrans** (Ratzeburg)
—   Mandible with lower tooth obviously longer (about 1·4 times) than upper (Fig. 99). Tergite 2 of gaster with diagonal furrows strong, subtending an angle of about 60 degrees to anterior margin. (Small to moderately large species, fore wing length 4·1–10·5 mm. Ovipositor–hind tibia index 4·4–5·1. Pterostigma usually yellowish with margins dark.) . . . . **agnoscendus** (Roman)

7   Pterostigma with at least anterior margin broadly blackish. Gaster tergite 1 with central area weakly transversely rugulose. (Hind tarsus segment 5 (measured dorsally to apex of lobes) usually slightly longer than segment 3. Large to very large species, fore wing length 7·3–19·0 mm. Ovipositor–hind tibia index 4·4–4·7.) . . . **tuberculatus** (Geoffroy)

—   Pterostigma yellowish with margins narrowly fuscous. Gaster tergite 1 with central area
     rugulo-punctate or punctate-reticulate. 8
8   Gaster tergite 1 with central area rugulo-punctate. Punctures on scutellum relatively fine.
     Hind tarsus segment 5 (measured dorsally to apex of lobes) usually slightly shorter than
     segment 3. (Large to very large species, fore wing length 8·2–20·5 mm. Ovipositor–hind
     tibia index 4·4–5·0. Male mid coxa with a large thorn-like projection on its outer side
     (Fig. 97).) . . . . **mesocentrus** (Gravenhorst)
—   Gaster tergite 1 with central area punctate-reticulate. Punctures on scutellum relatively
     strong (in comparison with *mesocentrus*). Hind tarsus segment 5 usually slightly longer
     than segment 3. (Large to very large species, fore wing length 7·8–20·9 mm. Ovipositor–
     hind tibia index 4·3–5·1.) . . . . **messor** (Gravenhorst)

D. agnoscendus. Rare but widely distributed, England: Cornwall, Devon, Hampshire,
Berkshire, Hertfordshire, Greater London, Kent, Norfolk; Wales: Dyfed; Ireland:
Kerry, Down. Flight period: v–vii. In addition to one specimen reared from the weevil
Mesites tardii (Curtis) we have seen a series (15) reared from the dead stems of *Rosa
canina* from which several species of beetles and small moths also emerged. We have
also seen 2 specimens reared from *Alnus* logs and have observed a female investigating
an *Alnus* log infested with the woodwasp *Xiphydria camelus* (Linnaeus), the woodwasp's pimpline parasitoids *Rhyssella* and *Pseudorhyssa*, and almost certainly other
insects as well. The specimens from *Rosa* are smaller and more slender than those
associated with *Alnus* and, although there is a continuous range of variation between
the extremes represented by these groups, more than one biological species may be
included in the segregate recognised here.

D. diversicostae. Rare (11 specimens from 5 localities), Scotland: Badenoch &
Strathspey and Kincardine & Deeside. Flight period: vi–viii. Reared from the
cerambycid beetle *Acanthocinus aedilis* (Linnaeus) in *Pinus sylvestris* (2 specimens;
Perkins, 1946).

D. imperator. Rare in England (Hampshire); widespread but still uncommon in
Scotland, from Ettrick & Lauderdale in the south to Sutherland in the north. Flight
period: vi–vii. In Scotland it is often collected in the relict Caledonian pine forest, but it
also occurs regularly in the ancient birch-dominated deciduous woods of the extreme
west and north, in the complete absence of pine. There is circumstantial evidence that it
attacks beetles feeding in long-dead timber, such as *Rhagium* species, in which case
marked host–plant associations would not be expected. However, we have seen one
specimen extracted from the borings of *Arhopalus rusticus* (Linnaeus), a cerambycid
beetle that is usually found in recently dead *Pinus*.

D. mesocentrus. Uncommon; widely distributed in the southern half of Great Britain,
as far west as Devon and north to Gwynedd, Greater Manchester and Humberside.
Flight period: v–ix. It appears to be associated principally with beetle hosts in dead oak
and beech but we have seen no reared material.

D. messor. Rare (1 specimen from 1 locality and 7 without data), England: Hereford &
Worcester. Flight period: ix. We have seen one specimen supposedly reared from the
sesiid moth *Synanthedon vespiformis* (Linnaeus).

D. populneus. Uncommon; widely distributed in southern England, as far west and
north as Hereford & Worcester and Norfolk, with a single record for Cumbria. Flight
period: v, overwintering as a prepupa, but the Cumbria specimen was captured in early
vii. Reared from the cerambycid beetle *Saperda populnea* (Linnaeus) (12 specimens),
and less often from the moths *Synanthedon flaviventris* (Staudinger) (2 and 2 from

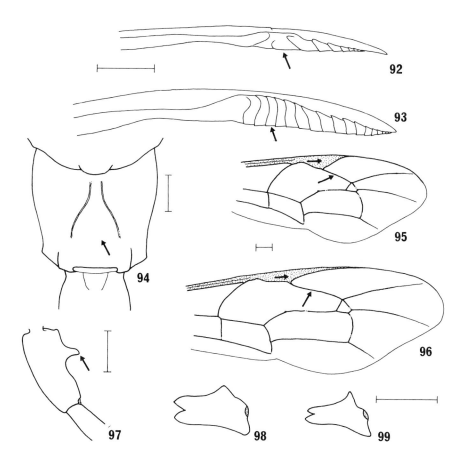

Figs 92–99. *Dolichomitus* species. 92, apex of ovipositor, lateral view, *D. imperator*. 93, apex of ovipositor, lateral view, *D. tuberculatus*. 94, propodeum, dorsal view, *D. diversicostae*. 95, part of right fore wing, *D. terebrans*. 96, part of right fore wing, *D. tuberculatus*. 97, left mid coxa, antero-ventral view, *D. mesocentrus* male. 98, left mandible, front view, *D. terebrans*. 99, left mandible, front view, *D.agnoscendus*. All scale lines represent 0·5 mm.

'*Salix* twigs') (Sesiidae) and *Lampronia fuscatella* (Tengström) (2) (Incurvariidae). These hosts all cause galls in small branches and twigs, respectively of *Populus* (mainly), *Salix* and *Betula*.

D. pterelas. Rare (10 specimens from 7 localities and 1 without data), England: Hereford & Worcester, Wiltshire, Oxfordshire, Essex, Suffolk; Ireland: Kerry. Flight period: vi–vii. Reared from the cerambycid beetle *Stenostola ferrea* (Schrank) (1 specimen).

D. terebrans. Uncommon but widely distributed among conifers in England, Scotland and Wales. Flight period: v–vi, viii; perhaps partly bivoltine throughout, and it overwinters as a prepupa. It has often been reared as a parasitoid of the large curculionid bark beetles attacking *Pinus* in Britain: *Pissodes castaneus* (Degeer) (15 specimens) and *P. pini* (Linnaeus) (2); and it has become a similarly regular (Evans, 1985) parasitoid of the recently established scolytid *Dendroctonus micans* Kugelann (2) attacking *Picea*.

D. tuberculatus. Uncommon but widely distributed in Great Britain and Ireland. Flight period: v–ix; probably bivoltine throughout. Reared from the bark weevil *Hylobius abietis* (Linnaeus) (14 specimens from one survey of this host), the cerambycids *Acanthocinus aedilis* (Linnaeus) (1) and ?*Rhagium mordax* (Degeer) (1), and more doubtfully from the sesiid moth *Synanthedon culiciformis* (Linnaeus) (1). Rearings have been from conifers and *Betula*.

Genus **TOWNESIA** Ozols

This genus includes only one species, *Townesia tenuiventris*, which could be accommodated in *Liotryphon*. *T. tenuiventris* has been associated particularly with *Liotryphon ruficollis*, on account of similarities in the shape of the gaster and the reduced submetapleural carina (Townes & Townes, 1960). However, partly for reasons of nomenclatural stability, we have maintained the generally accepted classification (Oehlke, 1967; Aubert, 1969; Townes, 1969; Kasparyan, 1981).

— Moderate to large species, fore wing length 7·4–12·9 mm. Body black with hind corner of pronotum yellow-marked, legs mainly red. Female with gaster tergite 3 1·8–2·0 times as long as broad; ovipositor–hind tibia index 5·1–6·0. Male with extremely long hairs on veins of the leading, proximal part of fore wing (including M + Cu). . .   **tenuiventris** (Holmgren)

T. tenuiventris. Rare, but widely distributed in southern England as far west as Somerset and north to Cheshire and Norfolk. Flight period: vi–viii. We have seen no reared British material, but have been able to examine one of the specimens from Jussila & Käpylä's (1975) study, in which they reared *T. tenuiventris* as a regular parasitoid of the megachiline bee *Chelostoma florisomne* (Linnaeus) and less often from the sphecid wasp *Trypoxylon figulus* (Linnaeus) nesting in dead wood at one site in Finland. We do not concur with Jussila & Käpylä (1975) in their presumption that the published host records genuinely indicate an unusually catholic host range: however, some specimens of *T. tenuiventris* are so large that either considerably larger hosts than *Chelostoma* and *Trypoxylon* must be used as well, or else it must behave as a pseudoparasitoid, consuming successive hosts in adjacent cells of wood-nesting aculeate Hymenoptera.

Genus **PARAPERITHOUS** Haupt

A small genus which, like *Townesia*, is closely associated with *Liotryphon*. There is one species in Europe and a few in the eastern Palaearctic (Townes, 1969).

—   Large species, fore wing length 10·3–14·2 mm. Body black, though gaster may have a rufescent tinge, legs mainly red. Ovipositor–hind tibia index 4·6–4·9; upper valve of ovipositor with a row of dorsolateral fine teeth. . . . . . .   **gnathaulax** (Thomson)

P. gnathaulax. Rare but widely distributed; England: Cornwall, Surrey, Essex; Scotland: North East Fife, Perth & Kinross, Kincardine & Deeside, Nairn, Aberdeen, Inverness. Flight period: vi–vii, ix–x. We have seen no reared material. The structure of the ovipositor suggests that it probes for hosts pupating beneath partly loose bark.

Genus **LIOTRYPHON** Ashmead

This is a moderately large genus comprising a rather diverse group of species. It is difficult to define and in many ways it can be considered as the group of long-ovipositored ephialtines left after segregation of all the genera characterised by distinctive autapomorphies. *Liotryphon* has been confused with *Dolichomitus* and placement of some species is rather arbitrary (see notes on *Dolichomitus* and *Townesia* above). In general *Liotryphon* species have weaker ovipositors than most *Dolichomitus* species, and they appear to use the ovipositor to probe bark crevices etc. or to penetrate the weak, prepared exit sites of insects with non-mandibulate adults (e.g. Lepidoptera) pupating in bark and wood, rather than to drill through bark. As presently constituted the genus extends across the Holarctic and into the Oriental and Neotropical regions. About 10 species are found in Europe and 5 occur in Britain.

Key to species

1   Submetapleural carina distinct only anteriorly (Fig. 100). Thorax (including pronotum, mesoscutum, mesopleuron, scutellum and postscutellum) with conspicuous red and cream marks. (Propodeum in profile sloping very gradually to posterior margin; conspicuously and relatively evenly punctate. Fore and mid coxae pale yellowish, hind coxa reddish. Fore wing length 7·2–8·6 mm. Ovipositor–hind tibia index 4·3–5·0.)   .   .   .   .
     .   .   .   .   **ruficollis** (Desvignes)
—   Submetapleural carina complete, reaching almost to metacoxa (Fig. 101). Thorax almost entirely black (or dark brown in some small or immature specimens), at most with cream mark on upper hind corner of pronotum and lower part of mesopleuron reddish. (Propodeum in profile sloping more strongly to posterior margin; less conspicuously and/ or less evenly punctate. In females all coxae more or less uniformly reddish or blackish.)
       2
2   Female.   3
—   Male.   6
3   Ovipositor with upper valve rounded at apex (Fig. 104). (Fore wing length 3·6–8·5 mm. Ovipositor–hind tibia index 4·1–4·6.)   .   .   .   .   **caudatus** (Ratzeburg) (female)
—   Ovipositor with upper valve simply tapered at apex (see Fig. 119).   4
4   Ovipositor–hind tibia index at least 6·0. Hind coxa blackish or dark brown. Gaster tergite 2 with obvious, impressed oblique furrows cutting off anterolateral corners. Body rather slender. (Fore wing length 5·0–6·8 mm.)   .   .   .   .   **strobilellae** (Linnaeus) (female)
—   Ovipositor–hind tibia index less than 5·0. Hind coxa reddish. Gaster tergite 2 with vestigial oblique furrows.   5
5   Legs more slender (Fig. 106); hind tibia about 0·33 as long as fore wing, hind tarsus segment 2 at least 5·7 times as long as broad. Upper hind corner of pronotum with a distinct cream line. Hair on ovipositor sheath relatively fine (a very good character when both species are available for comparison). (Fore wing length 4·9–10·0 mm. Ovipositor–hind tibia index 4·5–4·9.)   .   .   .   .   **punctulatus** (Ratzeburg) (female)
—   Legs less slender (Fig. 105); hind tibia about 0·30 as long as fore wing, hind tarsus segment 2 at most 5·2 times as long as broad. Upper hind corner of pronotum usually with only a cream or reddish spot. Hair on ovipositor sheath relatively coarse. (Fore wing length 5·0–10·9 mm. Ovipositor–hind tibia index 4·5–5·0.)   .   .   **crassisetus** (Thomson) (female)

41

6 Upper hind corner of pronotum with a distinct cream line or triangular spot which reaches almost to level of notaulus. Parameres elongate (Fig. 102). (Face more closely punctate, and therefore relatively dull and with more abundant silvery pubescence. Anterior tergites of gaster with strong punctate–reticulate sculpture.) 7

— Upper hind corner of pronotum with only a cream or reddish spot. Parameres short (Fig. 103). (Face more sparsely punctate, and therefore shining and with less abundant silvery pubescence.) 8

7 (Legs less slender, see Fig. 105 of female *crassisetus*. NOTE: this character works reasonably well when several specimens are available for comparison but absolute ranges of measurements overlap. Antennal scape entirely yellow ventrally.) . . . .
. . . . **caudatus** (Ratzeburg) (male)

— (Legs more slender, see Fig. 106 of female. Antennal scape sometimes mainly brown ventrally.) . . . . **punctulatus** (Ratzeburg) (male)

8 Mesothorax with median sternal groove effaced anteriorly, traceable only as a weakly sculptured line (Fig. 107). Gaster very elongate; shining; the anterior tergites with obvious deep, but quite widely spaced, punctures. . . . . . . **strobilellae** (Linnaeus) (male)

— Mesothorax with median sternal groove complete, extending forwards to epicnemial carina. Gaster not abnormally elongate; the anterior tergites with strong punctate–reticulate sculpture. . . . . **crassisetus** (Thomson) (male)

L. caudatus. Uncommon but widely distributed in southern England, west and north to Hampshire and Cheshire. Flight period: v–xi; presumably bivoltine, overwintering as a prepupa. We have seen numerous specimens reared from cocoons of the codling moth *Cydia pomonella* (Linnaeus) (Tortricidae) collected from apple bark in France and Cyprus, but in Britain only from the cocoon of ?*Pammene regiana* (Zeller) (Tortricidae) (1 specimen) collected under *Acer pseudoplatanus* bark. Several examples have been collected from the trunks of other angiosperm trees, and the structure of the ovipositor tip (Fig. 104) suggests that all of its hosts will be reached by probing the gaps in flaking, crevassed or loose bark.

L. crassisetus. Uncommon but widely distributed in southern England and as far north as Cumbria, and found in Ireland: Kerry and Cork. Flight period: v–ix, probably bivoltine (but see below). The principal hosts are sesiids and other Lepidoptera feeding in twigs, twig galls or the bark of mature trees, but the cocoons of other Lepidoptera found in similar situations seem to be attacked when encountered. Reared from ?*Zeuzera pyrina* (Linnaeus) (1 specimen) (Cossidae), *Synanthedon flaviventris* (Staudinger) (20 and 5 from 'Salix twigs') in the gall of which it overwinters, *S. myopaeformis* (Borkhausen) (4 and 1 from '*Malus* bark'), *S. scoliaeformis* (Borkhausen) (1), *S. culiciformis* (Linnaeus) (1) (Sesiidae), *Blastodacna atra* (Haworth) (4) (Momphidae), *Cydia pomonella* (Linnaeus) (5) (Tortricidae) and ?*Lycia hirtaria* (Clerck) (1) (Geometridae). A few of the specimens from *S. flaviventris* are extremely small, suggesting that they may have developed on first rather than second year hosts. *L. crassisetus* as here recognised is unusually variable, especially in regard to the length of the tergites and temples and the extent to which the latter are expanded behind the eyes. Specimens from *S. flaviventris* in *Salix* twigs generally fall near to the maximum development of these character states, in contrast to those from hosts such as *S. myopaeformis* in *Malus* bark, but many intermediates occur, especially in the non-reared material that we have seen. Detailed biological investigation would be needed to decide whether more than one species is involved, each with a narrower host range, or whether the morphological variation is associated with a seasonal alternation of hosts.

L. punctulatus. Rare (6 specimens from 5 localities); England: Hereford & Worcester, Cheshire, Oxfordshire, Berkshire, Greater London. Flight period: v–vi, viii. We have seen several Swiss specimens reared from the tortricid moth *Cydia funebrana* (Treitschke).

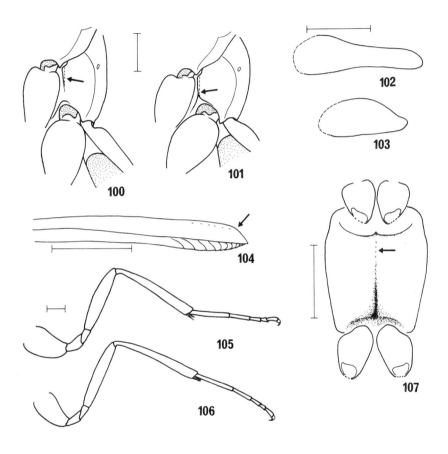

Figs 100–107. *Liotryphon* species. 100, metapleuron, ventro-lateral view, *L. ruficollis*. 101, meta-pleuron, ventro-lateral view, *L. caudatus*. 102, left paramere, lateral view, *L. punctulatus*. 103, left paramere, lateral view, *L. crassisetus*. 104, apex of ovipositor, lateral view, *L. caudatus*. 105, left hind leg, lateral view, *L. crassisetus* female. 106, left hind leg, lateral view, *L. punctulatus* female. 107, mesothorax, ventral view, *L. strobilellae* male. All scale lines represent 0·5 mm.

L. ruficollis. Rare (9 specimens from 8 localities and 4 without data); England: Somerset, Hampshire, Surrey, Greater London, Oxfordshire, Suffolk, Staffordshire. Flight period: vi–viii. We have seen no reared material but specimens have been collected high in mature oak trees.

L. strobilellae. Only one British record (Bridgman, 1886, as *Ephialtes strobilorum*): both sexes reared supposedly by Bignell, v–vi from '*Coccyx strobilorum*' (no doubt the tortricid moth *Cydia strobilella* (Linnaeus)); locality not stated but apparently not Devon as the species was not included in Bignell's (1898) Devon list. Bridgman (1886) clearly believed the specimens to be British; but possibly in error as Bignell usually supplied locality data for his own collecting activities, and received specimens from a variety of other sources. In addition to the above specimens we have seen material reared in Norway and Sweden from *Cydia strobilella* in *Picea* cones.

Genus **EXERISTES** Foerster

This small genus includes four European and one Nearctic species. In Britain there is only one species, *E. ruficollis*; which is anomalous within the genus because the tarsal claws of females lack a tooth-like basal lobe. This has been considered by some authors (e.g. Oehlke, 1967) sufficient reason to place it in a genus of its own, *Eremochila*.

—   Fore wing length 3·9–7·9 mm. Ovipositor–hind tibia index 3·9–4·3. Female black, with mesothorax, legs and generally most of gaster red brown. Male black, with gaster often rufescent, legs reddish and trochanters yellow. . . . . . .   **ruficollis** (Gravenhorst)

E. ruficollis. Uncommon but widely distributed among *Pinus* in mainland southern Britain as far north as Gwynedd, with Scottish populations centred on the native *Pinus sylvestris* woods of Badenoch & Strathspey and Kincardine & Deeside. Flight period: vi–vii; apparently univoltine and certainly overwintering as an adult to attack its tortricid hosts in iii–iv. An important parasitoid of the pine shoot moth *Rhyacionia buoliana* (Denis & Schiffermüller) in southern Britain (Thorpe & Caudle, 1938) and imported to Canada (Juillet, 1959) against this pest where, however, it failed to become established. In Scotland it is an abundant parasitoid of *Petrova resinella* (Linnaeus) on which it develops in broods of 1–3. We have also seen material reared from *Blastesthia turionella* (Linnaeus) (1 specimen) and associated with *Cydia coniferana* (Ratzeburg) (1), and presumably most of the tortricids whose feeding on *Pinus* (and perhaps other Pinaceae) result in resinous exudations will be included in the host range. Thorpe & Caudle (1938) suggested that *E. ruficollis* is thelytokous in error, having misidentified the male.

Genus **AFREPHIALTES** Benoit

A moderately small genus with most species in Africa and south-east Asia. Only one occurs in Europe. Townes (1969) treated this genus as a synonym of *Flavopimpla* but Gupta & Tikar (1969) have demonstrated that the two are distinct. The European species clearly belongs to *Afrephialtes*, to which it has been transferred by Kasparyan (1981).

—   Fore wing length 9·3–11·7 mm. Ovipositor–hind tibia index 3·9–4·1. Body black, with clypeus and sometimes gaster rufescent, hind corner of pronotum and tegula yellowish; legs mainly reddish. . . . .   **cicatricosa** (Ratzeburg)

A. cicatricosa. Rare (3 specimens from 2 localities); England: Dorset. The specimens seen have all been reared in vi: from the sesiid moth *Synanthedon formicaeformis*

(Esper) (2 specimens), and from a sallow branch containing a sesiid thought to be *S. formicaeformis* (1).

Genus **FREDEGUNDA** gen. nov.

Type species: *Pimpla diluta* Ratzeburg, 1852.

Diagnosis: Occipital carina mediodorsally dipped, complete. Clypeus with apical margin medially impressed, making it weakly bilobed. Fore wing with vein 3rs-m present. Hind wing with abscissa of Cu between M + Cu and cu-a longer than cu-a. Submetapleural carina present. Propodeum with lateromedian longitudinal carinae long and well developed. Ovipositor strongly compressed, its apex with lower valves enlarged and partially enclosing the upper valve (Fig. 68). Apical segment of tarsi swollen, tarsal claws long and large (Fig. 59). Female fore tarsal claw with a tooth-like basal lobe. Male with gaster tergite 1 narrowed towards posterior end (Fig. 85).

Etymology: From the proper name Fredegund, a wicked medieval queen who sent assassins in the guise of messengers (Gregory of Tours, *History of the Franks* 8: 44, cited in Leighton, 1972).

Gender: feminine.

This genus includes only the type species. Our principal reason for recognising it as a genus is to maintain equivalence among the genera in the tribe Ephialtini (see p. 48).

— Thorax and propodeum predominantly reddish; pronotum with a yellow stripe along upper hind margin. Head black. Remainder of insect mainly reddish. Fore wing length 6·0–9·0 mm. Ovipositor–hind tibia index 1·5–1·6. Male with fore femur slightly flattened ventrally. . . . . **diluta** (Ratzeburg) comb. nov.

F. diluta. Rare, restricted to reed beds and perhaps only in eastern England: from Sussex to Humberside. Flight period: vi–viii; probably overwintering as an adult. Reared in broods of 1–2 from the larvae of noctuid moths in *Phragmites* stems: *Archanara dissoluta* (Treitschke) (2 specimens, possibly from one host), *Arenostola phragmitidis* (Hübner) (1), and a series of parasitised hosts in which *Archanara* sp. and *Arenostola phragmitidis* were both present (20; includes 5 broods of 2). [We are convinced that the host datum for the specimen in BMNH reared by C. G. Barrett (see Morley, 1908: 68) is in error and, because we have seen no other material from the same area, we are also treating the locality information as erroneous. It is labelled as reared from the nymphalid butterfly *Cynthia cardui* (Linnaeus) in Wales: Dyfed.]

Genus **ENDROMOPODA** Hellén stat. nov.

Endromopoda is a medium-sized Holarctic genus. Townes (1969) treated the group as a subgenus of *Scambus*, but morphologically it is far more distinct from *Scambus* species than are, for example, some species of *Gregopimpla*. *Endromopoda* species share a number of specialised features, including having a strongly laterally compressed ovipositor with the apical teeth perpendicular to the ovipositor axis; a long propodeum; swollen distal tarsal segments; and bevelled tarsal claws (the latter two characters less well-developed in males). Unlike most species of *Scambus*, the female subgenital plate in *Endromopoda* is more or less uniformly sclerotised. As far as is known all species attack stem inhabiting or gall making hosts associated with monocotyledons, especially the larger species of Graminae. *Endromopoda* is represented in Britain by six species and no additional ones are known from continental Europe.

Two species of *Scambus, S. nigricans* and *S. cincticarpus*, are, in some features, somewhat intermediate between typical *Scambus* and *Endromopoda*. They similarly attack stem inhabiting hosts, and have the ovipositor compressed with the subgenital plate of the female fairly uniformly sclerotised but, unlike *Endromopoda*, they have strongly oblique apical teeth on the ovipositor and do not have the tarsal segments or claws modified.

Key to species (females only)

1   Ovipositor–hind tibia index 0·8–0·9; ovipositor in profile slightly up-curved; lower valve apically with about 5 teeth (Fig. 108). Gaster usually with tergites 2–6 red or reddish, rarely almost black; hind tibia uniformly reddish. Propodeum dorsally coarsely punctate, submatt. (Fore wing length 5·0–8·0 mm. Male fore femur without a ventral excavation.) . . . . **arundinator** (Fabricius)
—   Ovipositor–hind tibia index 1·1–1·6; ovipositor in profile more or less straight; lower valve apically with 7 or more teeth. 2
2   Gaster with tergite 3 sparsely and very unevenly punctate (Figs 109, 110). 3
—   Gaster with tergite 3 closely and fairly evenly punctate. 4
3   Tergites 3 and 4 red, each with a shallow transverse furrow along its anterior margin, which has close punctures arranged along it, and remainder of tergite more or less impunctate (Fig. 109). (Fore wing length 6·0–8·0 mm. Ovipositor–hind tibia index 1·5–1·6. Coxae red. Male fore femur with a deep long excavation on ventral side (Fig. 112).) . . . . **nitida** (Brauns)
—   Tergites 3 and 4 reddish-black, without a distinct transverse furrow anteriorly, and with only widely scattered punctures over their anterior 0·5 (Fig. 110). . . . . . . **species A**
4   Hind tibia and first tarsal segment reddish, both at most only weakly infuscate near apex. Ovipositor–hind tibia index 1·4–1·6. (Fore wing length 6·0–8·0 mm. Coxae reddish. Male fore femur with a double excavation on ventral side (Fig. 111).)   **phragmitidis** (Perkins)
—   Hind tibia dorsally with centre whitish, and with weak to strong subproximal and distal blackish marks; first tarsal segment white, distally infuscate. Ovipositor–hind tibia index 1·1–1·3. 5
5   Hind coxa black. Head behind the eyes rather strongly rounded and, in dorsal view, with ocelli forming a more or less equilateral triangle. (Fore wing length 3·0–6·0 mm. Male fore femur with a long shallow ventral excavation and hind coxa always black.) . . . . **nigricoxis** (Ulbricht)
—   Hind coxa nearly always red. Head behind the eyes weakly rounded, in dorsal view, with ocelli forming a rather broad-based triangle. (Fore wing length 4·0–8·0 mm. Male fore femur with a long shallow ventral excavation and hind coxa usually at least partly red except in undersized specimens.) . . . . **detrita** (Holmgren)

E. arundinator. Restricted to reed beds, where it can be locally common. England: Isle of Wight, Hampshire, Suffolk, Norfolk, Cambridgeshire; Ireland: Wicklow. Flight period: vi–ix. We have seen male specimens collected in June while swarming on a *Phragmites* stem from which a female was about to emerge, and one female collected ovipositing into a *Phragmites* stem at the end of July, but no material with more precise biological data.

E. detrita. Generally common and widespread in fertile grassy habitats throughout the British Isles. Flight period: v–ix; probably partly bivoltine at least in the south, over-wintering as a prepupa. Well known (Salt, 1931) as a minor parasitoid of the wheat stem-sawfly *Cephus pygmeus* (Linnaeus), but it attacks a wide range of other hosts inhabiting the stems of (or galling) many large species of Gramineae and even *Juncus*. We have seen material reared from *Cephus nigrinus* Thomson (2 specimens), *C. pygmeus* (96, all from concerted surveys of this host), *Calameuta filiformis* (Eversmann) (8) (Hymenoptera: Cephidae), *Tetramesa airae* (von Schlechtendal) (11) (Hymenoptera: Eurytomidae), *Mesoligia literosa* (Haworth) (1), *Archanara* sp. (3), *Coenobia rufa* (Haworth) (2) (Lepidoptera: Noctuidae), and indet. chloropid sp. (1) (Diptera). Probably less indicative of true hosts are single specimens labelled 'ex buoliana' (= the tortricid moth *Rhyacionia buoliana* (Denis & Schiffermüller)) and 'ex strigularia' (= the geometrid moth *Ectropis bistortata* (Goeze)), the latter being mounted with a ?campoplegine ichneumonid cocoon that looks much too small to have been the real host.

E. nigricoxis. Perkins (1943) and Oehlke (1967) synonymised *nigricoxis* with *detrita*. However, it forms a clear morphological segregate and, even in the absence of biological information, we feel justified in regarding it as a distinct species.

Uncommon, but widely distributed in grassland habitats throughout the British Isles, particularly in the north. Flight period: v–ix, but probably univoltine in the north. There are no host records.

E. nitida. Rare (11 specimens from 3 localities), restricted to fens; England: Norfolk. Flight period: vi–viii. There are no host records.

E. phragmitidis. Rare, restricted to reed beds; England: Hampshire, Cambridgeshire, Norfolk, North Yorkshire; Wales: South Glamorgan. Flight period: v–viii, overwintering as a prepupa. We have seen it reared from the galls of *Lipara* species (10 specimens) (Diptera: Chloropidae), in some cases having pupated in the looser plant tissue above the gall, and also from a dipterous host (possibly the predatory scatophagid *Cleigastra apicalis* (Meigen)) in the upper part of a frass filled but ungalled *Phragmites* stem (1). Chvála *et al.* (1974) suggested that in the Netherlands it is almost

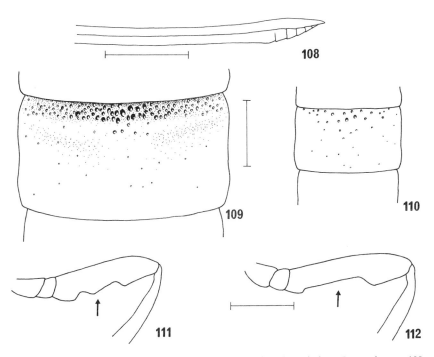

Figs 108–112. *Endromopoda* species. 108, apex of ovipositor, lateral view, *E. arundinator*. 109, tergite 3 of gaster, dorsal view, *E. nitida* female. 110, tergite 3 of gaster, dorsal view, *E.* species A female. 111, left fore femur, anterior view, *E. phragmitidis* male. 112, left fore femur, anterior view, *E. nitida* male. Scale lines represent 0·5 mm.

exclusively parasitic on *Lipara rufitarsis* (Loew) rather than *L. lucens* Meigen, but in some of its British localities only the latter seems to occur. Biological information given by Raghi-Atri (1980) under the name *Scambus detrita* may really refer to this species.

E. species A. We have not been able to identify this species. It may be new, but we are refraining from formally describing it because we have seen only one specimen.
   A single female; England: Norfolk, Barton Turf, in *Phragmites/Cladium* fen, 12–13.vii.1983 [in NMS].

Genus **SCAMBUS** Hartig

Scambus is a large genus that is represented by a diverse assemblage of species in the Holarctic region. A few species occur in South America and in montane habitats along the Oriental/Eastern Palaearctic interface. The concept of *Scambus* as adopted here differs from that of other recent authors (e.g. Oehlke, 1967; Townes, 1969; Aubert, 1969) in that we exclude species placed in the subgenus *Endromopoda* (see above), and place *diluta* in a separate new genus. *Scambus*, as here defined, is a far less heterogenous assemblage of species than the species grouped together by Townes (1969). Even so, it may be paraphyletic with respect to genera such as *Acropimpla* and *Gregopimpla*. These taxa could be considered to be species-groups of *Scambus*. The generic limits of *Scambus*, like those of other paraphyletic assemblages, are very difficult to define. Townes (1969) separated it from other genera with a medio-dorsally dipped occipital carina (except *Acropimpla*) on the grounds that in the hind wing the abscissa of Cu between M + Cu and cu-a is longer than cu-a (i.e. 'nervellus intercepted below the middle'). However, in Britain this is also the case in some specimens of *Gregopimpla* and the division between *Gregopimpla* and *Scambus* seems to be rather contrived. Likewise, in Britain the dividing line between *Scambus* and *Acropimpla* (see Townes, 1969: 67) seems to us to be rather artificial. Although the majority of *Acropimpla* species (which occur in the Old World tropics) do comprise a distinctive species group, several Palaearctic taxa are morphologically intermediate. The sole British *Acropimpla*, *A. didyma*, is such a species and it is questionable whether it is really phylogenetically more closely related to exotic *Acropimpla* or to some species included in *Scambus*.
   The species we here include in *Scambus* have often been placed in two subgenera, *S. (Scambus)* and *S. (Ateleophadnus)* (Oehlke, 1967; Townes, 1969) but we have chosen not to recognise these as their adoption would be inconsistent with the overall classification of European ephialtines. The two British *Scambus* which would be placed in *Ateleophadnus* are *S. nigricans* and *S. cincticarpus*. Both these species resemble *Endromopoda* in some respects (as noted under that genus).
   Between 17 and 23 species of *Scambus* have been said to occur in Europe, and 14 are here recognised as occurring in Britain. However, the limits of some species are very poorly defined. Several of the European putative species have been separated only on the very superficial basis of colour differences (see Aubert, 1966; 1967) and their status remains questionable in view of the known colour variation in several species. (For example, large females of *S. brevicornis* have red hind coxae, whilst small females have them black.) In some other cases, however, morphological variation within the species limits recognised here may eventually turn out to represent more than a single biological species.
   In trying to identify males it should be borne in mind that the coloration of the pterostigma is rarely as stable or as characteristic as in females. The colour of coxae is also a less useful character for males, and in species showing red hind coxae in the female sex the males, especially if small, often have them brown or blackish.

Key to species (females only)

1   Ovipositor–hind tibia index at most 1·9 (if 1·8–1·9 then malar space 0·3–0·5 times basal width of mandible and pterostigma centrally blackish or dark brown). (Malar space 0·3–0·5 times basal width of mandible. Subgenital plate with sclerotised hind margin about 0·4 length of sternite. Black: coxae red, though sometimes with fore coxa partly or entirely black. Propodeum finely punctate.) 2

—   Ovipositor–hind tibia index more than 2·2, or if 1·9–2·1 then with malar space 0·2 times basal width of mandible or with pterostigma pale yellowish. (Malar space and pterostigma otherwise various.) 3

2   Pterostigma centrally blackish or dark brown. Ovipositor–hind tibia index 1·5–1·9. Ovipositor at most very feebly downcurved. (Fore wing length 4·0–10·0 mm. Male fore femur strongly excavate ventrally.) . . . . **nigricans** (Thomson)

—   Pterostigma centrally yellowish. Ovipositor–hind tibia index 1·4–1·5. Ovipositor appreciably and evenly downcurved. (Fore wing length 4·9–8·9 mm. Male fore femur weakly excavate ventrally.) . . . . **cincticarpus** (Kriechbaumer)

3   Thorax and propodeum black but with mesoscutum, scutellum, mesopleuron and often metapleuron red marked. 4

—   Thorax and propodeum more or less unicolorous black or dark red brown. 5

4   Ovipositor–hind tibia index 2·8–3·1. Propodeum longer, and in profile evenly rounded (Fig. 113). (Fore wing length 3·0–7·0 mm. Pterostigma yellowish. Male fore femur with a weak double concavity ventrally. Male antenna with tyloids on segments 4–6.) . . . . **elegans** (Woldstedt)

—   Ovipositor–hind tibia index 2·0–2·3. Propodeum shorter, and abruptly declivous (Fig. 114). (Fore wing length 4·0–6·0 mm. Pterostigma yellowish. Male fore femur slightly deplanate ventrally. Male antenna without tyloids.) . . . . **pomorum** (Ratzeburg)

5   Hind tarsus with first segment 0·88–0·98 times as long as remaining tarsal segments (but excluding claws); upper hind corner of pronotum yellow, usually with a narrow stripe extending forwards along upper margin. 6

—   Hind tarsus with first segment 0·70–0·85 times as long as remaining tarsal segments (but excluding claws); upper hind corner of pronotum black or yellow, rarely with a stripe extending forwards along upper margin. 8

6   Head, in dorsal view, with genae strongly constricted behind the eyes (Fig. 116). Fore wing with pterostigma black. . . . . **species A**

—   Head, in dorsal view, with genae not or only weakly constricted behind the eyes (Figs 117, 118). Fore wing with pterostigma centrally pale translucent yellowish 7

7   Head in dorsal view somewhat buccate, with genae broad (Fig. 117). Ovipositor–hind tibia index 2·4–2·9. (Fore wing length 5·0–9·0 mm. Male fore femur with a pair of concavities ventrally.) . . . . **planatus** (Hartig)

—   Head in dorsal view not buccate, with genae evenly rounded. (Fig. 118). Ovipositor–hind tibia index 2·9–3·7. (Fore wing length 4·0–8·0 mm. Male fore femur with a pair of concavities ventrally.) . . . . **calobatus** (Gravenhorst)

8   Ovipositor–hind tibia index equal to or greater than 2·5 AND fore wing with pterostigma dark brownish, proximally and distally pallid, centrally more or less concolorous with Sc+R+Rs.. 9

—   Ovipositor–hind tibia index generally less than 2·6, if rarely 2·6–2·7 then pterostigma is pale yellowish centrally; pterostigma usually pale yellowish with anterior margin infuscate, but centrally paler than Sc+R+Rs, or if centrally dark then with ovipositor–hind tibia index less than 2·4. 10

9   Mid and hind coxae black or blackish. Nodus of ovipositor weak; upper valve not apically concave (Fig. 119). (Fore wing length 4·0–8·0 mm. Ovipositor–hind tibia index 2·6–3·5. Male fore femur flattened ventrally.). . . . . **sagax** (Hartig)

—   Mid and hind coxae reddish. Nodus of ovipositor strong; upper valve slightly concave between nodus and apex (Fig. 120). (Fore wing length 4·0–8·0 mm. Ovipositor–hind tibia index 2·5–3·1. Propodeum short, abruptly declivous (more so than in Fig. 114). Male fore femur with a double concavity ventrally.) . . . . **buolianae** (Hartig)

10   Pterostigma, when viewed against a dark background, with an obscure whitish band along posterior margin. Hind tarsus with first segment 0·64–0·74 times as long as remaining tarsal segments (excluding claws). (Coxae black, or large specimens with hind coxa

reddish marked. Fore wing length 3·0–8·0 mm. Ovipositor–hind tibia index 2·1–2·5. Male with ventral surface of fore femur weakly flattened, especially in small specimens. Male pterostigma often uniformly dark brownish.) . . . . **brevicornis** (Gravenhorst)

— Pterostigma, when viewed against a dark background, with posterior margin slightly infuscate, or concolorous with centre. Hind tarsus with first segment 0·76–0·83 times as long as remaining tarsal segments (excluding claws). 11

11 Apex of ovipositor, in profile, with upper valve abruptly constricted, but with a narrow apical section (Fig. 121). Orbital-ocellar distance virtually equal to inter-ocellar distance. First segment of hind tarsus white, only its distal 0·2 blackish. (All coxae reddish. Fore wing length 4·0–7·0 mm. Ovipositor–hind tibia index 2·0–2·2. Male with ventral surface of fore femur weakly flattened.) . . . . **foliae** (Cushman)

— Apex of ovipositor, in profile, with upper valve fairly evenly tapered to distal end (Figs 122, 123). Orbital ocellar distance 1·1 or more times as long as inter-ocellar distance. First segment of hind tarsus proximally whitish, but generally with distal 0·5 infuscate. . 12

12 Upper hind corner of pronotum black. Hind tibia reddish, with a blackish area at distal dorsal apex, rarely with an indistinct black mark near proximal end. Tegula with posterior 0·3 or more black. (Fore wing length 4·0–7·0 mm. Ovipositor–hind tibia index 2·2–2·4. Fore coxa often black or blackish. Flagellum generally entirely black proximally. Male with ventral surface of fore femur with a broad shallow concavity.). . . . .

. . . . **eucosmidarum** (Perkins)

— Upper hind corner of pronotum usually narrowly yellow marked and/or dorsum of hind tibia centrally and proximally whitish, with distal apex and a subproximal band black. Tegula usually entirely yellow, rarely with extreme posterior margin infuscate. . . 13

13 Propodeum in profile very evenly declivous, without distinct dorsal and posterior faces (Fig. 115). Female with subgenital plate uniformly sclerotised. (Fore wing length 3·5–9·4 mm. Ovipositor–hind tibia index 1·9–2·3.) . . **Gregopimpla inquisitor** (Scopoli) (see p. 56)

— Propodeum in profile abruptly declivous, with distinct dorsal and posterior faces (see Fig. 114). Female with subgenital plate somewhat membranous centrally. 14

14 Upper valve of ovipositor with a very strongly raised nodus; penultimate proximal tooth of lower valve angled at less than 45 degrees to axis of ovipositor (Fig. 122). (Fore wing length 3·0–7·0 mm. Ovipositor–hind tibia index 2·2–2·4. Male fore femur flattened ventrally.) . . . . **vesicarius** (Ratzeburg)

— Upper valve of ovipositor with nodus less conspicuous; penultimate proximal tooth of lower valve angled at about 45 degrees or more to shaft of ovipositor (Fig. 123). (Fore wing length 3·0–8·0 mm. Ovipositor–hind tibia index 2·0–2·7. Male fore femur with a double concavity ventrally.) . . . . **annulatus** (Kiss)

S. annulatus. Perkins (1943) added *S. signatus* (Pfeffer) to the British list on the basis, we presume, of two specimens now in the BMNH. These belong to the segregate here called *annulatus* and although the name *signatus* has priority we are refraining from using it until the taxonomy and nomenclature of *annulatus* has been further clarified (see below).

Common, widely distributed in the British Isles north to Argyll & Bute and Badenoch & Strathspey. Flight period: iv–x; bivoltine, overwintering as a prepupa. It is a regular parasitoid of leaf mining and leaf rolling Lepidoptera on deciduous trees and, as a result of intensive study of parasitism of Gracillariidae in particular, we are able to give many records from these groups (see also Askew & Shaw, 1986). However, specimens with a tendency towards longer temples have been reared regularly during surveys of Tortricidae feeding in pods of Leguminosae in the field layer, and the existence of a substantial proportion of specimens larger than any of the above in non-reared series also suggests that either we are not in a position to express the host range of this single species adequately or, more probably, that *S. annulatus* as here recognised is an aggregate of several biological species. We have seen it reared from the following Lepidoptera: *Eriocrania* sp. (1 specimen) (Eriocraniidae), *Tischeria ekebladella* (Bjerkander) (2), *T. angusticolella* (Duponchel) (2) (Tischeriidae), *Caloptilia elongella* (Linnaeus) (16), *C. betulicola* (Hering) (9), *C. rufipennella* (Hübner) (3), *C. alchimiella* (Scopoli) (31), *C. stigmatella* (Fabricius) (6), *C. robustella* Jäckh (3), *C. syringella*

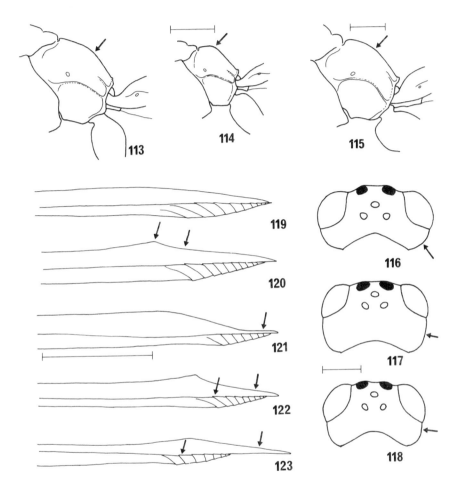

Figs 113–123. *Scambus* species and *Gregopimpla inquisitor.* 113, propodeum, lateral view, *S. elegans* female. 114, propodeum, lateral view, *S. pomorum* female. 115, propodeum, lateral view, *Gregopimpla inquisitor* female. 116 head, dorsal view, *S.* species A female. 117, head, dorsal view, *S. planatus* female, 118, head, dorsal view, *S. calobatus* female. 119, apex of ovipositor, lateral view, *S. sagax.* 120, apex of ovipositor, lateral view, *S. buolianae.* 121, apex of ovipositor, lateral view, *S. foliae,* 122, apex of ovipositor, lateral view, *S. vesicarius.* 123, apex of ovipositor, lateral view of *S. annulatus.* All scale lines represent 0·5 mm.

(Fabricius) (20), *Parornix anglicella* (Stainton) (1), *P. devoniella* (Stainton) (1), *P. betulae* (Stainton) (5), *P. scoticella* (Stainton) (1), *P. torquillella* (Zeller) (5), *Callisto denticulella* (Thunberg) (3), *Phyllonorycter quercifoliella* (Zeller) (3), *P. corylifoliella* (Hübner) (2), *P. salicicolella* (Sircom) (1), *P. spinolella* (Duponchel) (1), *P. maestingella* (Müller) (1), *P. coryli* (Nicelli) (1), *P. lautella* (Zeller) (1), *P. ulmifoliella* (Hübner) (2), *P. nicellii* (Stainton) (4), *P. trifasciella* (Haworth) (4), *Phyllonorycter* spp. (5) (Gracillariidae), *Atemelia torquatella* (Zeller) (2) (Yponomeutidae), *Coleophora virgaureae* Stainton (1) (Coleophoridae), *Perittia obscurepunctella* (Stainton) (1) (Elachistidae), *Diurnea fagella* (Denis & Schiffermüller) (2), *Agonopterix nervosa* (Haworth) (1) (Oecophoridae), *Mompha conturbatella* (Hübner) (10) (Momphidae), *Cydia succedana* (Denis & Schiffermüller) (2), *C.* ?*nigricana* (Fabricius) (15), *Epiblema roborana* (Denis & Schiffermüller) (1), *Epinotia immundana* (Fischer von Röslerstamm) (1), *Pandemis cerasana* (Hübner) (1), *Clepsis senecionana* (Hübner) (1), *Acleris schalleriana* (Linnaeus) (2) (Tortricidae), and *Stenoptilia pneumonanthes* (Büttner) (1) (Pterophoridae); additionally as a pseudohyperparasite from cocoons of *Apanteles* (sensu lato)/*Caloptilia* (17), *Macrocentrus*/indet. tortricid (1), *Rhysipolis*/*Caloptilia* (8) (Braconidae), indet. campoplegine/*Eucosma* (1), and *Phytodietus*/indet. tortricid (1) (Ichneumonidae). Also reared from leaf-mines of other Orders: *Zeugophora subspinosa* (Fabricius) (1) (Coleoptera: Chrysomelidae), *Orchestes salicis* (Linnaeus) (4) (Coleoptera: Curculionidae), *Scolioneura betuleti* (Klug) (1) (Hymenoptera: Tenthredinidae) (but on the basis of extensive sampling it is clear that this group is outside its normal host range), and once from an undetermined leaf-rolling weevil on birch. Single specimens reared from the stem galls of the sesiid moth *Synanthedon flaviventris* (Staudinger) and the tenthredinid sawfly *Euura amerinae* (Linnaeus) indicate that gallicolous hosts are sometimes attacked, though infrequently.

S. brevicornis. Widely distributed and common throughout the British Isles. Flight period: iv–ix; bivoltine, overwintering as a prepupa. It regularly parasitises concealed hosts feeding in prominent situations in the field layer, particularly in or on flower heads. Many substrate rearings from flower and seed heads of Compositae, especially, are unexpressed in the list that follows. Reared from the following Lepidoptera: *Caloptilia cuculipennella* (Hübner) (1 specimen), *Aspilapteryx tringipennella* (Zeller) (1) (Gracillariidae), *Glyphipterix haworthana* (Stephens) (3) (Glyphipterigidae), *Agonopterix putridella* (Denis & Schiffermüller) (5) (Oecophoridae), *Coleophora conspicuella* Zeller (5), *C. virgaureae* Stainton (1), *C. alticolella* Zeller (15, and 11 more from unspecified *Coleophora* spp. on *Juncus*), *C. adjunctella* Hodgkinson (2), *Coleophora* spp. on *Atriplex* (3) (Coleophoridae), *Metzneria metzneriella* (Stainton) (1), *Ptocheuusa paupella* (Zeller) (1), *Aristotelia brizella* (Treitschke) (1), *Platyedra subcinerea* (Haworth) (2), *Pexicopia malvella* (Hübner) (2), *Scrobipalpa salinella* (Zeller) (1), *Reuttia subocellea* (Stephens) (2), *Aproaerema anthyllidella* (Hübner) (2) (Gelechiidae), *Mompha conturbatella* (Hübner) (1) (Momphidae), *Dichrorampha consortana* Stephens (1), *Cydia* ?*succedana* (Denis & Schiffermüller) (1), *C. gemmiferana* (Treitschke) (1), *Lathronympha strigana* (Fabricius) (1), *Rhyacionia buoliana* (Denis & Schiffermüller) (2), ?*Eucosma tripoliana* (Barrett) (7), *Epiblema uddmanniana* (Linnaeus) (1), *Lobesia littoralis* (Humphreys & Westwood) (3), *Endothenia gentianaeana* (Hübner) (3), *Cnephasia stephensiana* (Doubleday) (1), *Acleris hastiana* (Linnaeus) (16, from one survey of this host) (Tortricidae), *Stenodes alternana* (Stephens) (3), *S. straminea* (Haworth) (13), *Cochylis flaviciliana* (Westwood) (3), *C. atricapitana* (Stephens) (7) (Cochylidae), *Pima boisduvaliella* (Guenée) (3), ?*Rotruda saxicola* (Vaughan) (2), *R. carlinella* (Hinemann) (1) (Pyralidae), *Platyptilia gonodactyla* (Denis & Schiffermüller) (21) (Pterophoridae), *Eupithecia linariata* (Denis & Schiffermüller) (1) (Geometridae), and *Orthosia gracilis* (Denis & Schiffermüller) (1)

(Noctuidae) and, as a pseudohyperparasite, from cocoons of the braconid *Apanteles* (sensu lato) (1) and the ichneumonids *Phytodietus* sp. (1) and indet. Campopleginae (3), the latter extracted from *Centaurea* heads. Also seen from the tephritid flies *Chaetostomella onotrophes* (Loew) (1) and *Urophora jaceana* (Hering) (2), and recorded by Janzon (1982) as a parasitoid of another tephritid in Sweden. One specimen is labelled as reared from the sesiid moth *Synanthedon andraeniformis* (Laspeyres), but we regard this as highly improbable.

S. buolianae. Moderately common and widespread in England north to Cheshire, and found in Scotland: Moray. Flight period: v–ix; bivoltine, overwintering as a prepupa. Well known as a parasitoid of the pest tortricid moth *Rhyacionia buoliana* (Denis & Schiffermüller) feeding in the shoots of young *Pinus* species and introduced, apparently unsuccessfully, to Canada for its control (Carlson, 1979). It also regularly attacks hosts associated with prominent field layer plants, especially those feeding in or on the relatively hard flower or seed heads of Compositae, Dipsaceae and Caryophyllaceae. In addition it has been reared from Lepidoptera that feed in *Rosa* shoots (including one long series) and once supposedly from a fungus feeder. At first sight such a host range seems to be ecologically abnormally wide for a pimpline, but we have not been able to recognise morphological segregates and it may be significant that *R. buoliana* attacks *Pinus* shoots often while the trees are still small and essentially within the field layer. Many substrate rearings, notably from heads of *Centaurea* and *Dipsacus* species, are unrepresented in the list that follows. We have seen it reared from the following Lepidoptera: *Nemophora scabiosella* (Scopoli) (1 specimen) (Incurvariidae), *Morphophaga boleti* (Fabricius) (1) (Tineidae), *Coleophora lineolea* (Haworth) (1), *C. silenella* Herrich-Schäffer (1) (Coleophoridae), *Metzneria lappella* (Linnaeus) (3), *Apodia bifractella* (Duponchel) (2), *Ptocheuusa paupella* (Zeller) (1) (Gelechiidae), *Rhyacionia buoliana* (Denis & Schiffermüller) (6), *Endothenia gentianaeana* (Hübner) (1), *Croesia bergmanniana* (Linnaeus) (61, from one survey of this host) (Tortricidae), *Cnaemidophorus rhododactyla* (Denis & Schiffermüller) (1) (Pterophoridae), and ?*Hadena bicruris* (Hufnagel) (1) (Noctuidae).

S. calobatus. Rare (15 specimens from 8 localities and 4 without data); England: Dorset, Wiltshire, Oxfordshire, Greater London, Kent, Essex, Suffolk. Flight period: vi–ix. Reared from *Acrobasis tumidella* (Zincken) (2 specimens) and *A. consociella* (Hübner) (5), and as a pseudohyperparasite from indet. ichneumonid/*Acrobasis* (1). The larvae of the above pyralid moths feed in relatively voluminous spinnings on *Quercus*. None of the reared specimens seen had overwintered, and it seems possible that this may be accomplished by the adult stage.

S. cincticarpus. As noted by Perkins (1943) this species is closely related to *S. nigricans*, with which Oehlke (1967) and Aubert (1969) incorrectly synonymised it. The clear differences in morphology (see key) leave no doubt that it is a distinct species, even without the supporting difference in host associations.
    Rare (4 specimens from 2 localities); England: Surrey, Sussex. Flight period: vi (reared), ix. We have seen it reared from the cephid sawfly *Hartigia linearis* (Schrank) both in Britain (1 series of 3 specimens) and on the continent.

S. elegans. Uncommon but widely distributed in southern Britain north to Norfolk, Cheshire and Clwyd, and also found in Ireland. Flight period: iv–ix; bivoltine, overwintering as an adult. It parasitises hosts feeding in concealed situations in the vegetation of trees and shrubs, in particular evergreens. We have seen it reared from the flowers and seedpods of *Ulex* galled by the cecidomyiid fly *Asphondylia ulicis* Verrall, in which, however, we consider it probable that it was parasitising an invading lepidopteran, (7 specimens including the type series of *Pimpla ulicicida* Morley, and 3 more from '*Ulex* pods'), and from the Lepidoptera *Caloptilia cuculipennella* (Hübner) (29, from one survey of this host), *C. leucapennella* (Stephens) (1) (Gracillariidae), *Argyresthia laevigatella* Herrich-Schäffer (1) (Yponomeutidae), *Cydia funebrana* (Treitschke) (1), *Rhyacionia buoliana* (Denis & Schiffermüller) (6) (Tortricidae), and as a pseudohyperparasite from cocoons of the braconid *Apanteles* (sensu lato)/*Caloptilia* (3) and the ichneumonid *Diadegma*/*Caloptilia* (4).

S. eucosmidarum. Rare and probably restricted to moorland habitats in England: Devon, Greater Manchester, Lancashire, Cumbria; and Wales: Gwynedd; but widely distributed though still rather uncommon in Scotland. Flight period: v–viii; apparently at least partly bivoltine even in the north, overwintering as a prepupa. It parasitises Lepidoptera concealed in the rolled leaves, flowerheads etc. of tall field layer plants and bushes. We have seen it reared from ?*Glyphipterix haworthana* (Stephens) (1 specimen) (Glyphipterigidae), *Agonopterix* sp. (1) (Oecophoridae), *Cydia aurana* (Fabricius) (3), *Hedya atropunctana* (Zetterstedt) (2) (Tortricidae), and *Achlya flavicornis* (Linnaeus) (1) (Thyatiridae).

S. foliae. A single British record (one female); Scotland: Glasgow City, autumn 1984, MV light trap in suburban garden. The species was originally described as a parasitoid of the leaf mining tenthredinid sawfly *Heterarthrus nemoratus* (Fallén) (Cushman, 1938), and we have seen continental material reared from mines of *H. ochropoda* (Klug) (5 specimens) and *H. vagans* (Fallén) (4). According to Dowden (1941) it overwinters as a prepupa.

S. nigricans. Common, widely distributed in the British Isles north to Argyll & Bute, Stirling and East Lothian. Flight period: v–x; bivoltine, overwintering as a prepupa. It principally attacks hosts feeding or pupating in the stems of prominent field layer plants, but it also attacks hosts in leaf rolls etc. on such plants (especially as a pseudohyperparasite). We have seen it reared from the following Lepidoptera: *Lampronia morosa* Zeller (1 specimen) (Incurvariidae), *Depressaria daucella* (Denis & Schiffermüller) (4), *D. pastinacella* (Duponchel) (5) (Oecophoridae), ?*Scrobipalpa clintoni* Povolný (1) (Gelechiidae), *Mompha conturbatella* (Hübner) (1) (Momphidae), *Rhyacionia buoliana* (Denis & Schiffermüller) (1), *Epiblema scutulana* (Denis & Schiffermüller) (2), *E. cirsiana* (Zeller) (1), ?*Aphelia paleana* (Hübner) (1), *Cnephasia stephensiana* (Doubleday) (1), *Cnephasia* sp. (1) (Tortricidae), *Aethes williana* (Brahm) or *A. francillana* (Fabricius) (2), *Cochylis atricapitana* (Stephens) (2) (Cochylidae), *Myelois cribrella* (Hübner) (1) (Pyralidae), *Adaina microdactyla* (Hübner) (6) (Pterophoridae), and *Lycaena dispar batavus* (Oberthür) (1); additionally as a pseudo-hyperparasite from cocoons of the ichneumonid *Glypta*/*Aphelia* (15, from one survey). We have also seen single specimens supposedly from *Hartigia linearis* (Schrank) (Hymenoptera: Cephidae), *Lixus algirus* (Linnaeus) (Coleoptera: Curculionidae) and ?*Orellia falcata* (Scopoli) (Diptera: Tephritidae).

S. planatus. Rare; England: Devon, Hampshire, Oxfordshire, Surrey, Greater London, Essex, Hereford & Worcester, Cheshire. Flight period: v–vi according to data on specimens seen but, to attack the hosts recorded below, it must also be active in late summer or autumn; overwinters as a prepupa. Reared from fallen acorns harbouring larvae of the weevil *Curculio glandium* Marsham (3 specimens, with 2 more from *Curculio* sp. in acorns and one just labelled 'from acorn'), from a larva (possibly of the cephid sawfly *Janus femoratus* (Curtis)) in the hollowed and slightly swollen apex of a living oak twig (1), and from spindle berries infested by the larva of the pyralid moth

Alispa angustella (Hübner) (1). This latter specimen has the pterostigma much darker than in the others that we have examined.

S. pomorum. Uncommon but widely distributed in the British Isles, though not yet recorded from Wales. Flight period: iii–ix; perhaps univoltine and certainly overwintering as an adult. The only reared material seen has been from the apple blossom weevil *Anthonomus pomorum* (Linnaeus) (52 specimens; includes 41 from one survey of this host), of which it is a well-known parasitoid.

S. sagax. Locally common among well established conifers and widely distributed in England and Scotland, but not so far recorded from Wales or Ireland. Flight period: iv–xi; probably bivoltine throughout, overwintering as a prepupa or perhaps as a pupa. It attacks hosts feeding concealed in resinous galls, shoots or cones of conifers. We have seen it reared from the following tortricid moths: *Cydia millenniana* (Adamczewski) [= *deciduana* (Steuer) sensu Kloet & Hincks 1972] (5 specimens), *C. conicolana* (Heylaerts) (3), *Petrova resinella* (Linnaeus) (9), *Rhyacionia buoliana* (Denis & Schiffermüller) (18), and *Blastesthia turionella* (Linnaeus) (1). We have also seen it reared from the weevil *Pissodes validirostris* (Sahlberg) (2) and from the tenthredinid sawfly *Pristiphora ambigua* (Fallén) (4, and 3 more from *Pristiphora* spp. in sitka spruce buds).

S. vesicarius. Uncommon; widely distributed in the British Isles but not so far recorded from Wales. Flight period: v–ix; bivoltine, overwintering as a prepupa. It is a regular though rarely numerous parasitoid in the leaf and stem galls of tenthredinid sawflies on *Salix*, in which it attacks inquilines (and no doubt other parasitoids) as well as the gall maker (Carleton, 1939), but evidently its host range sometimes extends to Lepidoptera feeding concealed in the foliage of *Salix* (and perhaps on nearby or ecologically similar plants) as well. We have seen it reared from galls of *Euura amerinae* (Linnaeus) (14 specimens), *E. atra* (Jurine) (1), *Pontania arcticornis* Konow (2), *P. collactanea* (Foerster) (9), *P. dolichura* (Thomson) (2), *P. pedunculi* (Hartig) (1), *P. proxima* (Lepeletier) (23; includes 21 from one survey of this host), *P. viminalis* (Linnaeus) (3), and *Pontania* spp. (13), and also the Lepidoptera *Phyllonorycter viminiella* (Sircom) (1) (Gracillariidae), *Anacampsis* sp. (1) (Gelechiidae), and as a pseudohyperparasite from a cocoon of the braconid *Apanteles* (sensu lato)/*Caloptilia* (1). Less convincing are small specimens labelled as reared from the nest of a rat (1) and from the noctuid moth *Acronicta euphorbiae* (Denis & Schiffermüller) (1), and one specimen from the momphid moth *Mompha conturbatella* (Hübner) mounted with a cocoon that seems too small for the parasitoid specimen.

S. species A. We have not been able to identify this species. It may be new, but we are refraining from formally describing it because we have seen only one specimen.

A single female; Wales: Gwynedd, Anglesey, Llangristiolus, Malaise trap by hayfield, 7–27.viii.1982 [in NMS].

Genus **ACROPIMPLA** Townes

A moderately large genus with most species in the eastern part of the Palaearctic and in the Oriental region. The two species that occur in Europe are more like *Scambus* than most. They differ from *Scambus* in having the female subgenital plate more uniformly sclerotised, but this character can be hard to appreciate in dried specimens. Fortunately, the facial colour pattern of the single British species differentiates it from *Scambus*, which all have entirely black faces.

— Fore wing length 5·2–8·5 mm (an exceptional male 3·2 mm). Female: face black with a yellow triangular mark below each antennal socket; ovipositor–hind tibia index 2·2–2·7; coxae red. Male: face and clypeus entirely yellow.. . . . . **didyma** (Gravenhorst)

A. didyma. Rare; widely distributed in south-east England, and as far west as Wiltshire and Dyfed and north to Gwynedd, Cheshire and Norfolk. Flight period: vi–ix, probably bivoltine and overwintering as a prepupa. A gregarious parasitoid of large, flimsily cocooned lepidopterous prepupae or fresh pupae. We have seen material reared from the lasiocampid *Philudoria potatoria* (Linnaeus) (2 broods). Adolfsson (1984) gives some life-history information on *A. didyma* as an occasional parasitoid of the non-British lasiocampid *Dendrolimus pini* (Linnaeus) in Scandinavia.

Genus ISEROPUS Foerster

A small genus, mainly Holarctic in distribution but extending into the Neotropical region. One species occurs in Europe. The characters which have been used to separate *Gregopimpla* from *Iseropus* are rather weak but, as in most other similar cases, we have opted to maintain the status quo.

— Predominantly black with legs mainly reddish; fore coxa sometimes partly black; hind tibia banded: proximally and centrally whitish, distally and subproximally blackish (this banding more striking in males). Fore wing length 5·0–7·8 mm. Female: Ovipositor–hind tibia index 1·2–1·5; pterostigma black or dark brown, generally pale at extreme proximal corner. Male: face and clypeus entirely yellow. . . . .   **stercorator** (Fabricius)

I. stercorator. Uncommon but widely distributed, with records from England, Wales, Scotland and Ireland. Flight period: v–viii; bivoltine at least in the south and over-wintering as a prepupa. A gregarious parasitoid of flimsily cocooned lepidopterous prepupae or fresh pupae. We have seen material reared from the lasiocampid *Philudoria potatoria* (Linnaeus) (6 broods) and the lymantriid *Orgyia antiqua* (Linnaeus) (1), and emerged cocoons almost certainly belonging to this species in that of an *Acronicta* sp. (Noctuidae) (1). Adolfsson (1984) discusses *I. stercorator* as a parasitoid of the non-British lasiocampid *Dendrolimus pini* (Linnaeus) in Scandinavia.

Genus GREGOPIMPLA Momoi

A small genus, Palaearctic and Oriental in distribution. Three species occur in Europe, of which one is found in Britain.

— Predominantly black with legs mainly reddish and yellow; hind tibia banded: proximally and centrally whitish, distally and subproximally infuscate to blackish. Fore wing length 3·5–9·4 mm. Female: ovipositor–hind tibia index 1·9–2·3; pterostigma yellowish. Male: face and clypeus entirely black. . . . .   **inquisitor** (Scopoli)

G. inquisitor. Rare (9 specimens and 1 brood of 23 from 9 localities and 4 without data) but widespread; England: Hampshire, Surrey, Kent, Greater London, Hertfordshire, Suffolk, Oxfordshire, Buckinghamshire; Scotland: Badenoch & Strathspey. Flight period: v–vi and viii–xi; presumably bivoltine. Depending on host size, a solitary or gregarious parasitoid of a wide range of prepupae and fresh pupae of Lepidoptera concealed within flimsy cocoons. We have seen broods reared from cocoons of the lymantriids *Orgyia antiqua* (Linnaeus) (2 specimens) and *Dasychira fascelina* (Linnaeus) (1; brood of 23), an undetermined tortricid on *Betula* (1; solitary) and in Germany the notodontid *Clostera anastomosis* (Linnaeus) (2 specimens).

Genus TROMATOBIA Foerster

Tromatobia is Holarctic and Neotropic in distribution. The range of variation in size, sculpture and colour has led to the proposal of a large number of names for European species and the taxonomic problems in the genus are by no means satisfactorily resolved. We recognise 5 species in Britain. The larvae feed on successive eggs in more or less exposed spider egg sacs ('pseudoparasitism').

Key to species

1   Orbit black, entirely or with a small yellow or red mark on vertex. Face entirely black. Subalar prominence and dorsal margin of pronotum black. Lateromedian longitudinal carinae of propodeum strong and extending length of dorsal, horizontal part of propodeum (Fig. 124). 2
—   Orbit almost always narrowly to broadly yellow. Face centrally black, red or yellow. Subalar prominence and dorsal margin of pronotum yellow. Lateromedian longitudinal carinae of propodeum absent or, rarely, present anteriorly (Fig. 125). 3
2   Fore wing with vein 3rs-m present (although only faintly pigmented compared with 2rs-m). Orbit black with a small yellow to dark red spot on vertex (very rarely spot absent). Ovipositor tip more elongate (Fig. 126). Ovipositor–hind tibia index 0·9–1·1. Gaster often partly reddish. (Fore wing length 4·0–6·7 mm.) . . . . . **variabilis** (Holmgren)
—   Fore wing with vein 3rs-m absent entirely. Orbit entirely black. Ovipositor tip less elongate (Fig. 127). Ovipositor–hind tibia index 0·8–0·9. Gaster entirely black. (Fore wing length 3·7–5·4 mm.) . . . . **forsiusi** (Hellén)
3   Tergite 2 of gaster with relatively strong diagonal grooves cutting off antero-lateral corners and lateromedially much more sparsely punctate than rest of tergite. Segments of gaster often partly reddish, especially posteriorly. Corner of yellow orbital mark near hind ocellus making a sharp acute or right-angle. (Hind trochanter and trochantellus at least partly pale cream and with a pale spot at distal apex of femur (remainder reddish, sometimes with black marks). Puncturation of propodeum extending across the median longitudinal area, at least at its midlength. Male: face entirely yellow. Fore wing length 4·6–7·5 mm.) . . . . **oculatoria** (Fabricius)
—   Tergite 2 of gaster with at most weak diagonal grooves cutting off antero-lateral corners and relatively uniformly punctate. Segments of gaster usually uniformly black. Corner of orbital stripe near hind ocellus making a right angle or rounded. 4
4   Hind trochanter, trochantellus and femur uniformly reddish. Tergite 2 of gaster with very strong punctures (the posterior margin of most punctures as well defined as the anterior margin). Corner of orbital stripe near hind ocellus rounded. Puncturation of propodeum extending across the median longitudinal area, at least at its midlength. Male: face black with yellow orbital stripes. (Fore wing length 5·1–7·4 mm.) . **ornata** (Gravenhorst)
—   Hind trochanter and trochantellus at least partly pale cream and with a pale spot at distal apex of femur (remainder reddish) (occasionally all of these pale marks are rather limited and obscure). Tergite 2 of gaster with less strong punctures (the posterior margin of many

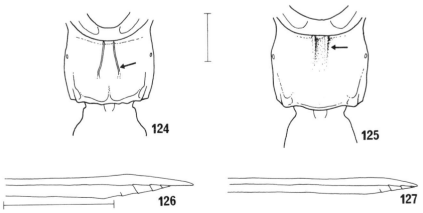

Figs 124–127. *Tromatobia* species. 124, propodeum, dorsal view, *T. variabilis*. 125, propodeum, dorsal view, *T. oculatoria*. 126, apex of ovipositor, lateral view, *T. variabilis*. 127, apex of ovipositor, lateral view, *T. forsiusi*. Scale lines represent 0·5 mm.

punctures much less well defined than the anterior margin). Corner of orbital stripe near hind ocellus making a right angle or rounded. Propodeum with the median longitudinal area largely impunctate, but sometimes less obviously smooth and shining at its mid-length. Male: face entirely yellow. (Fore wing length 4·5–7·4 mm.)   **ovivora** (Boheman)

T. forsiusi. Rare (3 specimens from 2 localities); Scotland: Badenoch & Strathspey, Kincardine & Deeside. Flight period: vi–vii. There are no host records.

T. oculatoria. Common throughout Britain and Ireland. Flight period: v–xi, in at least two generations even in the north. Reared in broods of c. 1–6 from the egg sacs of various spiders, especially relatively woolly ones constructed among the leaves of bushes and tall herbs and exposed on buildings. The following list is somewhat weighted towards species in which the female spider broods or stays near the egg sac, because only a small proportion of unattended sacs can be identified with confidence: *Philodromus aureolus* (Clerck) (3 broods), *Philodromus cespitum* (Walckenaer) (2), *Philodromus* sp. (1), *Tibellus oblongus* (Walckenaer) (3), *Tetragnatha* sp. (1), *Zygiella x-notata* (Clerck) (10), *Araneus diadematus* Clerck (4), *Araniella?cucurbitina* (Clerck) (4) and *Gongylidium rufipes* (Sundevall) (1). Female spiders guarding egg cocoons do not seem to be harmed by the ovipositing female parasitoid. Egg sacs quite often produce a few young spiders as well as the parasitoid adults. The winter is passed in overwintering egg sacs, in at least some cases as partly grown larvae which complete their feeding in early spring. Nielsen (1923) commented on the need for an alternation of hosts in his account of this species in Denmark.

T. ornata. A single British record: one female, Devon, Bovey, vii.1934. We have not seen reared material from Britain. On the continent it is regularly reared, gregariously, from *Argiope bruennichi* (Scopoli) egg sacs (C. Rollard, P. Sacher, pers. comms), but it seems unlikely to be restricted to this host.

T. ovivora. Although the taxonomy of the European species of *Tromatobia* remains in need of further investigation and clarification, the retention of *rufipleura* (Bignell) as a separate species cannot be justified. It falls within the range of variation of *ovivora* as here recognised, with which we therefore formally synonymise it. The type material of *rufipleura* has been examined (Fitton, 1976).
Uncommon but widespread in England, Scotland and Ireland, though not so far found in Wales. Flight period: v–x, in at least two generations even in the north. Reared in broods of 1–2 from the woolly egg sacs of *Araniella* sp. (3 broods) on *Pinus* and *Zygiella x-notata* (Clerck) (1) in a greenhouse. Spiderlings also emerged from one of the *Araniella* sacs.

T. variabilis. Rare, in dune slacks, heathland, etc.; England: Cornwall, Devon, Dorset, Hampshire, Surrey; Wales: Dyfed, Gwynedd; Ireland: Kerry, Kildare and 'Kilmore' (the type locality of *Pimpla hibernica* Morley). Flight period: vi–viii. Reared in broods of 1–2 from egg sacs of *Agalenatea redii* (Scopoli) (1 brood, and 2 more from egg sacs of similar bluish appearance), and *Larinioides cornutus* (Clerck) (1).

Genus **ZAGLYPTUS** Foerster

This small genus has a worldwide distribution and three species occur in Europe. Two are found in Britain. The larvae develop on mature spiders and their eggs, within egg nests.

Key to species

1   Scutellum usually red, always marked with yellow posteriorly; face yellow marked, below antennal sockets at least; hind tibia with black bands narrower than intermediate white area. Female: swellings on 2nd and 3rd tergites of gaster with mainly very fine punctures, contrasting with the strong punctures on the adjacent parts of the tergites. Male: antennal segments 6 and 7 with raised tyloids on outer, distal half of each (Fig. 128). (Fore wing length 4·3–6·0 mm. Mesothorax often reddish, gaster blackish. Ovipositor–hind tibia index 1·2–1·4.) . . . . **multicolor** (Gravenhorst)

—   Scutellum entirely black; face entirely black; hind tibia with black bands at least as broad as intermediate white area. Female: swellings on 2nd and 3rd tergites of gaster with punctures almost as strong as on adjacent parts of the tergites. Male: antennal segments 8 to 10 with tyloids, giving the appearance of notches between segments 8 and 9 and 9 and 10 (Fig. 129). (Fore wing length 3·0–5·5 mm. Body predominantly black, gaster sometimes partly reddish. Ovipositor–hind tibia index 1·3–1·5.). . . . . . **varipes** (Gravenhorst)

Z. multicolor. Uncommon, but widespread in southern England and as far north as Norfolk and Cheshire. Flight period: vi–ix. Although we have seen 2 broods (of 2 and 5 individuals) labelled as reared from the egg nests of *Enoplognatha ovata* (Clerck) in rolled *Rubus* leaves, circumstantial evidence suggests that its usual hosts may be arboreal.

Z. varipes. Uncommon, but widespread in Great Britain north to Badenoch & Strathspey, and also found in Ireland. Flight period: v–ix. We have seen 6 broods (of 2–6 individuals) reared from the egg nests of *Clubiona* spp. (mostly or perhaps all *reclusa* O. Pickard-Cambridge) in rolled *Rubus* and *Rumex* leaves. Nielsen (1935) gave an account of the biology of this species in Denmark, where he found it parasitising the egg nests of *Cheiracanthium erraticum* (Walckenaer) and, less frequently, *Sitticus floricola* (C. L. Koch). He stated that the female spider is stung to death and about 2 to 4 (exceptionally up to 8) eggs are laid in the nest: the resulting larvae develop on both the eggs and the dead adult, and can do so even if the female spider has not laid her eggs before being attacked.

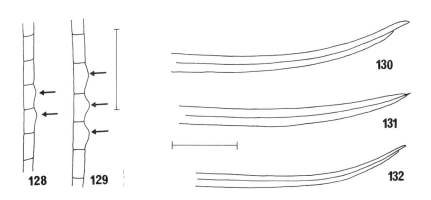

Figs 128–132. *Zaglyptus* and *Clistopyga* species. 128, segments 5–9 of right antenna (3–7 of flagellum), dorsal view, *Z. multicolor* male. 129, segments 7–11 of right antenna, dorsal view, *Z. varipes* male. 130, apex of ovipositor, lateral view, *C. rufator*. 131, apex of ovipositor, lateral view, *C. sauberi*. 132, apex of ovipositor, lateral view, *C. incitator*. Scale lines represent 0·5 mm.

Genus **CLISTOPYGA** Foerster

Clistopyga is a very widely distributed genus, although absent from Australia. There are 4 species in Europe, of which 3 occur in Britain.

Females are fairly easily recognised by the fairly long, upcurved ovipositor and large subgenital plate. Males are very distinctive because of the groove behind the malar space which is bounded by a high ridge. The larvae develop in concealed spider egg cocoons.

Key to species

1   Gastral tergites uniformly red, except for tergite 1 anteriorly (or rarely completely) and the posterior margins of tergites 3 and 4 (narrowly and sometimes only laterally) black. Thorax and propodeum black, with any reddish patches restricted to the ventro-lateral part of mesothorax. Male: face black, at most with very small yellowish orbital spots. Female: upper valve of ovipositor somewhat blunt ended (Fig. 130). (Fore wing length 4·5–6·2 mm.) . . . . **rufator** Holmgren
—   Gastral tergites black, sometimes partly reddish, especially laterally, but never uniformly so. Thorax and propodeum black, sometimes with extensive reddish marks. Male: face black or partly or entirely cream. Female: upper valve of ovipositor more gently tapering to apex (Figs. 131, 132). 2
2   Female. 3
—   Male. 4
3   Ovipositor very stout and relatively weakly upcurved (Fig. 131). Thorax and propodeum usually entirely black. (Antenna with 21–25 segments. Orbits black, at most with small yellow mark on vertex. Metapleuron at most punctate only posteriorly, punctured area not extending forward beyond level of propodeal spiracle. Fore wing length 5·0–5·5 mm.) . . . . **sauberi** Brauns (female)
—   Ovipositor slender and more strongly upcurved (Fig. 132). Thorax and propodeum usually, often extensively, yellow-marked and generally also red-marked. (Antenna with 24–32 segments. Orbits usually entirely cream. Metapleuron punctate posteriorly, the upper part at least to level of propodeal spiracle. Fore wing length 2·9–6·4 mm.) . . . . **incitator** (Fabricius) (female)
4   Face entirely (except sometimes for a narrow median vertical mark) cream. Clypeus and often entire orbit also cream. . . . . **incitator** (Fabricius) (male)
—   Face entirely or mainly black. Clypeus and orbit with restricted or no pale marks. . . . . **sauberi** Brauns (male)

C. incitator. Common and widely distributed throughout Britain and Ireland. Flight period: v–x, presumably bivoltine, at least in the south where adults peak around vi and viii–ix. Occurs particularly on tree trunks and walls. Very little is known for certain about the biology of this species, but abundant circumstantial evidence links it to spider egg sacs deeply concealed in holes and crevices. Nielsen (1929) recorded finding two larvae in a single egg sac of *Segestria senoculata* (Linnaeus) in a stone wall, one of which he succeeded in rearing.

C. rufator. Rare (16 specimens from 7 localities, and 3 specimens without locality data); England: Cambridgeshire, Norfolk, Somerset; Wales: Gwynedd. Flight period: vii–ix. Restricted to fens. Reared from the nest of *Clubiona juvenis* Simon (1 specimen) in curled dead *Phragmites* leaves: two elongate, frail, white cocoons (one destroyed by a predator) were present among what appeared to be the remains of consumed eggs, with the female spider dead in attendance.

C. sauberi. Townes & Townes (1960) suggested that this species was synonymous with the Nearctic *C. canadensis*. This synonymy was formally adopted by Aubert (1969) and has been applied subsequently by Kasparyan (1981). However, comparison of the limited material from Europe and North America in the BMNH collection suggests that it is better to use the European name for the European specimens, pending a more comprehensive study.

Rare (11 specimens from 5 localities); England: Cambridgeshire, Norfolk, Suffolk. Flight period: vii–ix. Apparently restricted to fens. There are no host records.

Tribe POLYSPHINCTINI

The Polysphinctini is a holophyletic group of genera whose larvae develop on mobile spiders. The adults are characterised by the possession of enlarged pulvilli, inflated tarsal segments and a relatively evenly tapered, very sharply pointed ovipositor. Most species also lack cross vein 3rs-m from the fore wing and many have strongly tapered mandibles and a smooth, highly polished cuticle. Worldwide the tribe comprises 13 genera and 9 of these occur in Europe. In Britain there are 19 species representing 7 genera. The two European genera which have not been found in Britain are *Oxyrrhexis* (which has a single Holarctic species and is closely related to *Polysphincta*) and *Zabrachypus* (a small Holarctic genus with one species in Europe).

The polysphinctines are probably the best known, and sometimes thought to be the only, ichneumonids associated with spiders. However, other pimplines (of the tribe Ephialtini), some members of the subfamily Phygadeuontinae and a range of other Hymenoptera also attack spiders and their eggs. This rather diverse assemblage has been reviewed recently by Fitton, Shaw & Austin (1987) and their paper includes a key to all the European genera (except Pompilidae) involved.

Townes & Townes (1960: 216–219) and Townes (1969: 96–98) traced the evolution of the Polysphinctini, parasitising mobile immature spiders, from ancestral habits of attacking spider egg cocoons, and in particular those that are in well defined nests guarded by the parent spider. Habits transitional to those of Polysphinctini are said to be present in extant ephialtine genera such as *Clistopyga* and *Zaglyptus*. These include attack on the guarding spider, sometimes before it has laid its eggs, and development on both adult and eggs (Nielsen, 1935), as well as consumption of spiderlings within the egg cocoon rather than the eggs themselves (Bignell, 1898). These suppositions are based on somewhat limited observations, however, and a careful reinvestigation of the biology of *Zaglyptus* and especially *Clistopyga* would be most valuable.

Detailed observations on the host-stinging and oviposition habits of adult Polysphinctini, in relation to their morphological specialisations, would also be of great interest — especially as the available accounts (Bignell, 1898; Nielsen, 1923; Cushman, 1926) are somewhat different and in any case rather incomplete. Polysphinctini are remarkable in being the only Pimplinae known to allow the host some further development after being attacked, and also for the extreme specialisation of their larvae. These are able to remain in position even through the host's moulting (though dead larvae are invariably sloughed when the host moults), and they fix themselves to a caked saddle of their progressively accumulated exuviae on the host integument by means of ventral protuberances (Nielsen, 1923). In addition, polysphinctine larvae in their final instars develop more or less paired wart-like dorsal protuberances on several consecutive body segments, each retractile and furnished with outward projecting hooks. These structures differ according to genus (Nielsen, 1923; 1928; 1935; 1937), and are used by the larva to grip the silk spun by the host as the parasitoid larva releases its hold on the host itself to finish its feeding, and then to move rather easily among strands of the host's silk as it constructs its cocoon.

In the tribe Ephialtini the final instar larvae of the genera *Zaglyptus* (Nielsen, 1923, 1935) and *Clistopyga* (Nielsen, 1929) are furnished with broadly similar hook-bearing warts. *Tromatobia*, which similarly feeds in spider egg cocoons but is arguably more primitive, has less elaborate structures bearing only unhooked spines (Nielsen, 1923).

Fields of cuticular hooks forming holdfasts also occur in the larvae of *Sericopimpla*, a mainly tropical genus that attacks Psychidae (Smithers, 1956). In all cases these structures seem to aid movement and orientation via the host's silk during cocoon formation, albeit in slightly different ways.

Although a number of bivoltine species seem to oviposit onto mature spiders for one of their generations, polysphinctines are particularly associated with immature spiders. In general the egg appears to hatch quickly but development of the parasitoid larva is at first very slight so that the immature host continues to feed and usually moults. Spiders with polysphinctine larvae on them are most often seen in autumn and spring, because so many species overwinter as minute larvae on active hosts.

The cocoons of the Polysphinctini are highly specialised structures, and in some genera so distinctive that brief descriptions are included in the sections below for those that we have seen. In some genera (*Acrodactyla, Sinarachna, Polysphincta, Zatypota*) they are constructed with a distinct opening at the caudal end, through which the prepupa defecates. In others (*Dreisbachia, Schizopyga*), and in the most closely related ephialtine genera, the opening is less clear and faecal material often adheres partly within the cocoon.

Most of the British polysphinctines are rather slender insects of inconspicuous appearance, generally blackish in colour with partly lighter coloured, and sometimes strongly patterned, legs. Attention is drawn to body colour only when it is other than blackish.

Genus **DREISBACHIA** Townes

A small genus with species widely scattered around the world, but only one occurs in Europe. The Holarctic species of *Dreisbachia* are immediately recognisable among Polysphinctini in having cross vein 3rs-m present in the fore wing.

— Fore wing length 3·9–5·9 mm. Eye with sparse but obvious short hairs. Mandibles strongly tapered with the upper tooth by far the longer. Ovipositor–hind tibia index 0·5–0·6. Ovipositor slightly upcurved. Male subgenital plate as in Fig. 89. Head and body blackish-brown with mesothorax red marked, especially at sides and below, and face variably yellow marked, at least with a weak spot below each antennal socket. .   **pictifrons** (Thomson)

D. pictifrons. Uncommon but widely distributed; England: Berkshire, Greater London, Norfolk; Scotland: Angus, Badenoch & Strathspey; Ireland: Wicklow. Flight period: vi–viii, x; possibly bivoltine in the south. Reared from *Clubiona* sp. or spp. (7 specimens): half-grown hosts found in spring each bore a small larva positioned on the side of the cephalothorax above the coxae, which had probably overwintered in that stage. Cocoon: translucent white, fusiform, closely woven and somewhat papery but expanded and frail, spun in the host's retreat. Circumstantial evidence suggests that *D. pictifrons* may be associated especially with semi-arboreal *Clubiona* species.

Genus **SCHIZOPYGA** Gravenhorst

Schizopyga is a relatively small Holarctic genus with four species in Europe, of which three have been found in Britain. Members of the genus can be separated from all other pimplines by their confluent, weakly convex face and clypeus, a feature found also in some members of the ichneumonid subfamilies Metopiinae and Orthocentrinae. The fourth European species, *S. varipes* Holmgren (also known as *S. flavifrons* Holmgren), has been recorded as British but we have seen no British specimens; it is easily identified by its entirely yellow face and clypeus.

1   Female. (Ovipositor–hind tibia index about 0·1).   2
—   Male.   4
2   Gaster red, typically with part of tergite 1, posterior of 2 and 3, and 6 onward entirely, black. Hind coxa usually red. (Fore wing length 3·6–4·6 mm. Mid femur usually red with distal apex black.). . . . .   **circulator** (Panzer) (female)
—   Gaster entirely black or blackish-brown. Hind coxa blackish.   3
3   Gaster narrower, tergite 2 at least 1·2 (usually more than 1·25) times as long as broad. Mid femur usually entirely red. (Fore wing length 3·5–7·3 mm.)   **frigida** Cresson (female)
—   Gaster wider, tergite 2 less than 1·2 (usually less than 1·15) times as long as broad. Mid femur red, but usually dark at distal apex. (Fore wing length 4·4–5·4 mm.) . . . .
. . . .   **podagrica** Gravenhorst (female)
4   Mid femur usually entirely red. Gaster with tergite 2 centrally longitudinally striate-coriaceous, the oblique grooves and transverse furrow more distinctly longitudinally striate; the area behind the transverse furrows on tergites 2 and 3 for the most part smooth and shining.. . . . .   **frigida** Cresson (male)
—   Mid femur red, usually with the distal apex black or blackish. Gaster with tergite 2 much more evenly coriaceous, sometimes with a few indistinct longitudinal striae; the area behind the transverse furrow on tergites 2 and 3 for the most part coriaceous. . . . . .
. . . .   **podagrica** Gravenhorst and **circulator** (Panzer) (males)

S. circulator. We have failed to find characters to separate the males of this species (in which the gaster is almost invariably completely black) and *S. podagrica*: it may be that we have not yet seen a male of *podagrica*.

Widely distributed in England, Scotland and Ireland (though we have seen no Welsh specimens); largely restricted to marshy habitats and uncommon in the south, but becoming commoner northwards. Flight period: iv–ix; probably bivoltine throughout. We have seen 10 specimens reared from *Clubiona* species determined as *terrestris* Westring (5), *?terrestris* (1), *neglecta* O. Pickard-Cambridge (1), *trivialis* C. L. Koch (1) and indet. (2). Some had certainly overwintered as larvae on subadult hosts. However, taken together with substrate rearings and its early time of appearance, the data associ-ated with others suggest that this species might sometimes overwinter as a fully fed larva, and that it oviposits in about v/vi and viii, perhaps sometimes onto adult hosts. The egg is placed at the hind margin of the cephalothorax. Cocoon: white, fusiform, rather thinly and loosely woven, expanded and frail, spun in the host's retreat.

S. frigida. Uncommon, but widely distributed in Britain, especially in the south but as far north as Inverness and Moray, and also found in Ireland: Down, Tipperary. Flight period: v–x, probably bivoltine, at least in the south. Seemingly a woodland species. Reared from *Clubiona terrestris* Westring (8 specimens) and *C. lutescens* Westring (1). Overwinters as a larva on subadult hosts. Egg placement and cocoon as described for *S. circulator.*

S. podagrica. At present males cannot be separated certainly from those of *circulator*. On the basis of females (4 specimens from 4 localities and 2 specimens without data) the species is rare and may be restricted to wetlands; England: Berkshire, Surrey; Scotland: Kincardine & Deeside. Flight period: v, viii–ix. We have seen no reared material, but Nielsen (1935) gives details of its life-history as a parasitoid of *Cheiracanthium erraticum* (Walckenaer) in Denmark.

Genus **PIOGASTER** Perkins

A small genus with three species in Europe, one in Japan and one in California. All are rare in collections and there are no host records. Circumstantial evidence suggests that some, at least, are arboreal. The tibiae of the two British species have conspicuous lengthwise pale and dark stripes.

The only males seen are believed to be *punctulata* and therefore we do not know if the key characters will work for males as well as females.

Key to species

1   Mesoscutum smooth and polished with scattered fine punctures, sparsely pubescent. Tergite 1 of gaster with lateromedian carinae not extending behind spiracle. (Fore wing length 3·9–5·0 mm. Ovipositor–hind tibia index about 0·9.) . . . . . **albina** Perkins
—   Mesoscutum coarsely and closely punctured, the punctures separated by less than their diameters, and densely pubescent. Tergite 1 of gaster with lateromedian carinae extending almost to hind margin. (Fore wing length 4·0–4·2 mm. Ovipositor–hind tibia index about 0·9.) . . . . **punctulata** Perkins

P. albina. Known to us only from one female; England: Norfolk, Kings Lynn, viii.1911; but also recorded from Leicestershire (Owen *et al.*, 1981).

P. punctulata. Rare (1 female and 1 male from 2 localities); England: Surrey, Greater London. Flight period: v–vi.

Genus **POLYSPHINCTA** Gravenhorst

Most of the pimpline species associated with spiders have at some time been referred to 'Polysphincta'. The name is now restricted to a rather small group, which is widely distributed in the northern hemisphere and South America. Five species occur in Europe, all of which are found in Britain. The genus includes some of the largest Palaearctic polysphinctines and is primarily associated with Araneidae.

Key to species

1   Scutellum, mandibles and subtegular ridge mainly yellow. Ovipositor–hind tibia index at least 1·3 OR gastral tergites strongly and densely punctured (the punctures separated by less than their diameter). (Submetapleural carina present and complete.) 2
—   Scutellum, mandibles and subtegular ridge black, red and black, or red. Ovipositor–hind tibia index less than 1·2 and gastral tergites less strongly punctured (the punctures on the sublateral rounded swellings separated on average by at least twice their diameter).   3
2   Ovipositor–hind tibia index 1·3–1·4. Gastral tergites entirely black, the punctures on the sublateral rounded swellings separated on average by at least twice their diameter. (Fore wing length 4·2–9·2 mm.) . . . . **boops** Tschek
—   Ovipositor–hind tibia index less than 1·0. Gastral tergites 2–5 or 6 each with the anterior margin pale (mainly whitish), the punctures on the sublateral rounded swellings separated by less than their diameter. (Fore wing length about 5·5 mm.) . . **nielseni** Roman
3   Submetapleural carina present as a very weak ridge or absent. Lateromedian longitudinal carinae of propodeum very weak or absent. Thorax and propodeum almost entirely black or extensively reddish. (Fore wing length 3·4–5·9 mm. Ovipositor–hind tibia index 0·8–1·0. Leg colour variable but hind coxa usually same colour as mesonotum. Mesoscutum and gastral tergites moderately hairy, more or less intermediate between the two species separated in couplet 4.) . . . . **tuberosa** Gravenhorst
—   Submetapleural carina complete, present as a distinct raised keel (see Fig. 87). Lateromedian longitudinal carinae of propodeum usually well defined anteriorly. Thorax and propodeum black. 4
4   All coxae and trochanters red (coxae rarely blackish on their inner surfaces). Mesoscutum with its anterior 0·7 evenly hairy. Gastral tergites, especially 2–4, with the punctures discernible among the coriaceous sculpture centrally and the tergites more or less uniformly hairy. (Fore wing length 3·7–7·2 mm. Ovipositor–hind tibia index 0·8–1·0.) . . . . **rufipes** Gravenhorst

— All coxae and trochanters black (the latter paler only at extreme apex). Mesoscutum with far fewer hairs on the middle 0·3 than on the anterior part, except along the notauli. Gastral tergites, especially 2–4, with very few punctures discernible among the rugulo-coriaceous sculpture centrally and the tergites with relatively fewer hairs. (Fore wing length 3·9– 5·8 mm. Ovipositor–hind tibia index 1·0–1·1.). . . . . **vexator** sp. nov.

P. boops. Rare (4 specimens from 3 localities), perhaps in long established woodland; England: Hampshire, Cumbria; Scotland: Perth & Kinross. Flight period: viii, and reared vi–vii. Reared from immature *Araniella* ?*cucurbitina* (Clerck) (2 specimens), on which the larva overwintered and was positioned dorsally and transversely at the base of the abdomen, and *A. opistographa* (Kulczynski) (1). Cocoon: diaphanous white, fusiform, with a very open and springy construction in which the coarse silk spans are straight and fastened in random directions to give an impression of more or less triangular arrangements.

P. nielseni. A single British record: England: Kent, Faversham, one female reared vii.1987 from an unidentified spider. We have seen an Austrian specimen reared from *Cyclosa conica* (Pallas) and Nielsen (1982) gave an account of *P. nielseni* as a parasitoid of this host. Cocoon: light brownish, fusiform, densely woven with a loose outer cover of coarser threads.

P. rufipes. Widespread, locally common in fens in England, Wales and Scotland, and also found in Ireland. Flight period: vi–ix. Reared from immature *Larinioides cornutus* (Clerck) (16 specimens), on which the larva overwinters and is positioned dorso-laterally and transversely at the base of the abdomen. In the Scottish Highlands it is univoltine, with adults emerging vi–vii and some larvae on very small hosts even overwintering a second year, but in southern Britain it appears to be bivoltine with emergences around vi and ix. Cocoon: as described for *P. boops*, but sometimes (? ageing) yellowish; always spun concealed in the host's retreat.

P. tuberosa. Common and widely distributed in England, Wales and Scotland; chiefly in bushy places but also on heathland. Flight period: vi–x. Appears to be univoltine in Scotland, emerging vi–vii, but at least partly bivoltine in the south, emerging vi and again viii–ix. Reared from immatures of *Araniella* ?*cucurbitina* (Clerck) (27 specimens, and 5 more from arboreal 'green spider'), *Araneus diadematus* Clerck (20), *Araneus quadratus* Clerck (3) and *Zygiella atrica* C. L. Koch (1). The larva is usually positioned as described for *P. boops*, and it overwinters on the host. Cocoon: as described for *P. boops*, but sometimes (?ageing) yellowish. Nielsen (1923) gave an account of the biology of this species in Denmark.

P. vexator sp. nov.
   Female: Fore wing length 3·9–5·8 mm. Mesoscutum with far fewer hairs on the middle 0·3 than on the anterior part, except along the notauli. Lateromedian longitudinal carinae of propodeum usually well defined anteriorly. Submetapleural carina strong and complete. Gastral tergites, especially 2–4, with very few punctures discernible among the rugulo-coriaceous sculpture centrally, the tergites otherwise polished and with relatively few hairs. Ovipositor–hind tibia index 1·0–1·1.
   Colour mainly black with legs partly reddish. Palpi and tegulae pale, but the tegulae often extensively infuscate. All coxae and trochanters black (the latter paler only at extreme apex); fore and mid femora, tibiae and tarsi mainly reddish; hind femur reddish, usually blackish apically; hind tibia infuscate, reddish on the upper surface and sometimes more extensively; hind tarsus infuscate with the proximal part of each segment reddish.

Male: Similar to female, usually with the legs darker and with the proximal part of the mid and hind tarsal segments pale (almost whitish) rather than reddish. This species is related to *P. tuberosa* and *P. rufipes*. In addition to the characters given in the key there are differences in the colour of the hind tibia: in *P. tuberosa* the upper surface is mainly pale (almost whitish) with a subproximal blackish band, whereas in the other two species it is usually reddish, more clearly so in *P. rufipes*.

HOLOTYPE female: ENGLAND: Cumbria, Foulshaw Moss, ex. *Araneus quadratus* coll. 30.v.87, em. 30.vi.87 (*Shaw*) (NMS).

PARATYPES, 9 females and 15 males, same data as holotype except dates (coll. 26.iv & 30.v.87, em. range 21.vi–10.vii.87) (NMS and BMNH); 1 female, WALES: Dyfed, Borth Bog, ex. *Larinioides cornutus* coll. 24.vii.85, cocoon 24.viii.85, em. 20.ix.85 (*Shaw*) (NMS); 1 female, FINLAND: Kuusamo, Nurmisaari, Paanajarvi, 2.viii.1935 (*Kerrich*) (BMNH); 1 female, SWEDEN: Stockholm, Sodra Ljustero, 27.viii.1933 (*Kerrich*) (BMNH).

P. vexator appears to be associated with grassy peat bogs and mosses, where its host is *Araneus quadratus* (Clerck) (25 specimens reared from immatures). In addition a single, particularly small specimen has been reared from *Larinioides cornutus* (Clerck). Cocoon: as described for *P. boops*.

Genus **ACRODACTYLA** Haliday

This is a moderate-sized genus with species in the Holarctic and through the Oriental region to Australia. Six species are found in Europe, of which three occur in Britain. At least one of the species-groups (the *quadrisculpta*-group: Townes, 1969) is sometimes accorded generic status (as *Colpomeria*), but in fact *madida* seems to occupy the more isolated position in *Acrodactyla*.

Key to species

1 Metapleuron rugose. Fore and mid femora subcentrally swollen, less than 3·8 times as long as maximally broad, lower side excised distally, usually with a tooth at base of excision (Figs 133, 134). (Fore wing length 2·8–5·0 mm. Ovipositor–hind tibia index 0·3–0·5. Vertical carina on mesoscutum in front of notauli well developed, obvious (Fig. 135).)
. . . . **quadrisculpta** Gravenhorst
— Metapleuron smooth with some fine weak puncturation. Fore and mid femora simple, symmetrical, more than 4·0 times as long as broad. 2
2 Vertical carina on mesoscutum in front of notauli well developed, obvious (see Fig. 135). Lateral lobes of mesoscutum with isolated hairs. (Fore wing length 2·8–4·4 mm. Ovipositor–hind tibia index 0·3–0·4.) . . . . **degener** (Haliday)
— Vertical carina on mesoscutum in front of notauli indistinct or absent. Lateral lobes of mesoscutum with dense fine pubescence. (Fore wing length 3·6–5·4 mm. Ovipositor–hind tibia index 0·3–0·4.) . . . . **madida** (Haliday)

A. degener. This morphospecies may comprise more than one biological species. Two segregates can be characterised as follows: lateral ocellus nearer to eye, posterior part of mesoscutum and gastral tergites 2–4 more hirsute, and segment 3 of hind tarsus longer; as opposed to: lateral ocellus further from eye, posterior part of mesoscutum and gastral tergites 2–4 less hirsute, and segment 3 of hind tarsus shorter. However, we presently consider the differences insufficient for the formal recognition of distinct species, nor are they supported by any biological or distributional data.

Probably the commonest British polysphinctine; widely distributed throughout Britain and Ireland. Particularly in woods, but also occurs in open habitats and even field crops. Flight period: vi–x; at least partly bivoltine, even in Scotland. All 58 reared specimens seen are from Linyphiidae, chiefly but not always immatures: ?*Bathyphantes* sp. (1), *Lepthyphantes* or *Bathyphantes* sp. (1), *Kaestneria dorsalis* (Wider) (1), *Boly-*

phantes alticeps (Sundevall) (1), *Lepthyphantes minutus* (Blackwall) (2), *Lepthyphantes tenuis* (Blackwall) (6), *Lepthyphantes zimmermanni* Bertkau (3), *Lepthyphantes mengei* Kulczynski (2), *Lepthyphantes* sp. (7), *Linyphia triangularis* (Clerck) (3), *Linyphia ?montana* (Clerck) (2), *Linyphia peltata* Wider (18, partly a sampling artefact due to disproportionate collecting from conifers), *Linyphia* sp. (7) and linyphiid (4). The larva is usually positioned dorsally or dorsolaterally on the basal half of the abdomen and it overwinters on the host. Cocoon: opaque whitish to deep brown (possibly depending on environmental conditions when constructed), spindle-shaped with a sharply ribbed, square cross-section, rather smooth and papery, usually constructed in the host's web. Nielsen (1923; cf. 1928) gave an account of the biology of this species in Denmark (at first misidentified as *Polysphincta pallipes* Holmgren), where he considered it to be thelytokous. Some British populations similarly seems to produce very few males, though in others males are abundant.

A. madida. Uncommon but widely distributed, with records from England, Scotland and Ireland. Flight period: v–ix; probably partly bivoltine, at least in the south. Found in wooded habitats. The usual hosts are immatures of *Metellina segmentata* (Clerck) and/or *M. mengei* (Blackwall) and perhaps *M. merianae* (Scopoli) (11 specimens) but we have also seen one specimen apparently reared from *Lepthyphantes* sp. The larva is positioned dorsally or dorsolaterally, near or basad of the middle of the abdomen. Unusually for a British polysphinctine, the winter is normally passed as a prepupa in a dense, brown rough cocoon that is variable in shape: subcylindrical or fusiform, often with the long axis curved and having some partly flattened sides. This is usually spun in ix or x, but hosts may sometimes go through at least part of the winter with well grown larvae on them. Nielsen (1923) gave an account of the biology of this species (as *Polysphincta clypeata* Holmgren) in Denmark.

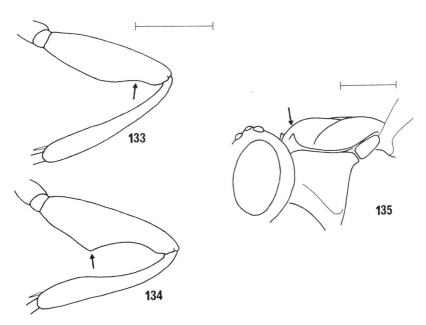

Figs 133–135. *Acrodactyla* species. 133 and 134, left mid femur, front view, *A. quadrisculpta*. 135, mesoscutum, antero-lateral view, *A.quadrisculpta*. Scale lines represent 0·5 mm.

A. quadrisculpta. The development of the tooth on the mid and fore femora varies considerably.

Uncommon; widely distributed in Britain and Ireland. Flight period: v–x; at least partly bivoltine, even in Scotland. Reared from immatures of *Tetragnatha obtusa* C. L. Koch (5 specimens), *T. extensa* (Linnaeus) (5), *T. ?montana* Simon (1) and *Tetragnatha* sp. (4), and in the midsummer generation from male and female adults of *T. extensa* (6). These hosts have been collected both from damp low vegetation and from trees and bushes. The larva is usually positioned towards the base of the abdomen, dorsally or dorsolaterally, and it overwinters on the host. Cocoon: like that of *A. degener*. Nielsen (1937) included notes on the biology of this species in Denmark.

Genus SINARACHNA Townes

The genus comprises seven species widely distributed in the northern hemisphere, of which four occur in Europe. Two species are here recorded from Britain. Another, species *S. anomala* (Holmgren), has also been recorded, apparently erroneously, from England. It can be distinguished from the other European species by its antenna of 16–18 segments (as opposed to 22 or more), and on the continent it has several times been recorded from *Dictyna* species.

Key to species

1   Tergites of gaster with more abundant, longer hairs (with numerous hairs antero-medially on segments 3 and 4). Antenna usually entirely blackish. Hind coxa usually mainly blackish. Female gaster more slender: tergite 1 1·4 or more times as long as broad. (Fore wing length 3·5–4·7 mm.) . . . . **nigricornis** (Holmgren)

—   Tergites of gaster with less abundant, shorter hairs (with segments 3 and 4 almost bare antero-medially). Proximal, ventral part of antenna usually pale or yellow. Hind coxa usually yellowish or reddish. Female gaster less slender: tergite 1 less than 1·3 times as long as broad. (Fore wing length 3·5–4·1 mm.) . . . . **pallipes** (Holmgren)

S. nigricornis. Uncommon but widely distributed; England: Hampshire, Surrey, Greater London, Cumbria; Scotland: Perth & Kinross, Kirkaldy, North East Fife, Midlothian, West Lothian. Flight period: vi–viii; probably univoltine, even in the south. Reared from immatures of *Araneus diadematus* Clerck (5 specimens), *Atea sturmi* (Hahn) (7) and *?Theridion* sp. (1) collected on trees and bushes. It seems particularly associated with *A. sturmi* on conifers, though only a small proportion of the larvae found have been reared successfully. The larva, which overwinters on the host, is positioned dorsolaterally and usually on the posterior half of the host's abdomen, with which it is aligned. Cocoon: light brownish, narrowly fusiform, and densely woven with a tight outer cover of coarse fibres; spun in the host's web.

S. pallipes. Rare; England: Oxfordshire, Hampshire, Surrey, West Sussex, Cambridgeshire, Cheshire, Cumbria. Flight period: vii–ix. Reared from *Araniella* sp. (2 specimens), and several times from unspecified spiders on deciduous trees. Cocoon: like that of *S. nigricornis*.

Genus ZATYPOTA Foerster

This is a moderately large genus, almost worldwide in distribution. All four European species occur in Britain. The species are small, often overlooked, and have an ovipositor–hind tibia index of less than 0·4. The hosts of this genus are regularly submature or even mature when killed. Although the hosts on which overwintering takes place are usually relatively small immatures, it seems possible that bivoltine species may attack adult hosts in the summer generation. However, direct observations are lacking.

Key to species

1 Hind wing with distal abscissa of Cu present though sometimes faint (Fig. 35). Mesoscutum coriaceous with pale hairs separated by about their own length. Propodeum with a shallow median longitudinal groove, the lateromedian longitudinal carinae indistinct. (Fore wing length 3·2–4·2 mm. Thorax, and sometimes gaster, usually extensively reddish or with whitish markings as well.) . . . . **bohemani** (Holmgren)
— Hind wing with distal abscissa of Cu entirely absent (see Fig. 36). Mesoscutum smooth and polished with isolated hairs. Propodeum almost always with the lateromedian longitudinal carinae strong. 2
2 Tergites of gaster smooth and polished. (Fore wing length 3·4–4·9 mm. Thorax and gaster usually partly reddish and with cream markings as well.) . . . **albicoxa** (Walker)
— Tergites of gaster with at least the raised central area coriaceous. 3
3 Metapleuron irregularly rugose. More robust species: antenna short, second flagellar segment less than 2·0 (female) or 2·5 (male) times as long as broad. (Fore wing length 3·4–4·4 mm.) . . . . **discolor** (Holmgren)
— Metapleuron indistinctly granulate, at most with short rugosities close to base of hind coxa. Less robust species: antenna longer, second flagellar segment more than 2·2 (female) or 2·7 (male) times as long as broad. (Fore wing length 2·8–4·6 mm. Thorax often extensively reddish, especially in southern specimens.). . . . . . . **percontatoria** (Müller)

Z. albicoxa. Rare (4 specimens from 3 localities); England: Hampshire, Oxfordshire. Flight period: vi–viii. Reared in Britain from *Achaearanea simulans* (Thorell) (1 specimen). We have also seen European material reared from this host (2) and from *A. lunata* (Clerck) (1). Cocoon: whitish to brownish, subcylindrical, rather densely spun with whorls of looser silk on its surface. Nielsen (1923) gave a detailed account of this species (as *Polysphincta eximia* Schmiedeknecht) as a parasitoid of *A. lunata* in Denmark.

Z. bohemani. Uncommon; widely distributed in England and as far north as the central lowlands of Scotland, but there are no records from Wales or Ireland. Flight period: v–x; probably largely bivoltine even in Scotland, with emergences v–vi and vii–viii. Occurring on tree trunks, walls and hedges. Reared from *Theridion mystaceum* L. Koch (3 specimens) and *T. ?mystaceum* (2). The larva is positioned more or less dorsolaterally, usually towards the base of the host's abdomen, where it overwinters. Cocoon: diaphanous, white (?ageing to pale yeollow), subcylindrical, with a springy and very open construction of sparse loose whorls, attached to the substrate at the wider capital end. The cocoons of this and the following two species are remarkably similar to those of certain Hemerobiidae (Neuroptera), and this no doubt accounts for the several literature records of polysphinctines being reared from hemerobiids.

Z. discolor. See note on *percontatoria* below.
Uncommon but widely distributed; England: Berkshire, Surrey, Cambridgeshire, Cumbria; Scotland: North East Fife, Kincardine & Deeside, Perth & Kinross. Flight period: vii–viii, x; probably univoltine in Scotland but perhaps bivoltine in the south. Found particularly in heathlands, often on conifers. Reared from *Theridion sisyphium* (Clerck) (10 specimens) and *T. sisyphium* or *T. impressum* L. Koch (2). The larva overwinters on the host and is positioned as described for *Z. bohemani*. Cocoon: like that of *Z. bohemani* but broader.

Z. percontatoria. We are using this name in a different sense from some other recent workers. Aubert (1969; 1970) and Kasparyan (1981) applied the name *percontatoria* to the species we here call *discolor* and the name *gracilis* to that which we call *percontatoria*. The neotype designation of Aubert (1969; 1970) does not comply with the requirements of the *International Code* and there are discrepancies in the identifications of type specimens made by Aubert (1970) and Šedivý (1963 and in Oehlke, 1967). Aubert (1970) also overlooked the question of the identity of *Pimpla phoenicea* Haliday, which (from the original description) must be the species he names as *gracilis* and of which it is a senior synonym. Pending solution of these nomenclatural problems, which are beyond the scope of this handbook, we are maintaining the usages and interpretations of Šedivý (1963) and Townes & Townes (1960). The confusion does not extend to separation of the species themselves; it only involves their names.

Moderately common; widely distributed in mainland Britain but not found in Ireland. Flight period: vi–x; mainly univoltine in the Scottish Highlands, but perhaps bivoltine elsewhere. Found especially on bushes. Reared from *Theridion varians* (25 specimens, 19 of which are from one locality), *T. simile* C. L. Koch (3), *T. tinctum* (Walckenaer) (2), and *Theridion* sp. (6). The larva overwinters on the host and is positioned as described for *Z. bohemani*. Cocoon: like that of *Z. bohemani*.

Tribe PIMPLINI

Adults of this tribe are characterised by having the mesopleural suture virtually straight and the abscissa of Cu between M + Cu and cu-a in the hind wing very short. The final instar larva is rather more distinctive in having a strongly developed epistomal arch and lacking a hypostoma. Adults have a rather robust appearance, with the ovipositor shorter than the gaster. The Pimplini includes some of the most common and conspicuous ichneumonids. Several species are quite large and attract attention in suburban gardens and similar situations.

The genera included in the Pimplini almost certainly constitute a holophyletic group. Their origin and relationship to other pimplines is, however, obscure. Nine genera are recognised worldwide. Some, such as *Pimpla* and *Xanthopimpla*, have large numbers of species. In Britain we have 3 genera represented in 19 species. The fourth European genus, *Strongylopsis*, is restricted to the Mediterranean area.

All three British genera are chiefly endoparasitoids of the pupae of Lepidoptera (*Itoplectis* species also having a marked association with ichneumonoid cocoons), but with different host-range foci. *Itoplectis* species are much more strongly associated with 'microlepidoptera' than with 'macrolepidoptera', while in *Pimpla* the principal host range emphasis shifts to 'macrolepidoptera'. *Apechthis* host ranges include both these groups, with butterflies central to some. Only *Pimpla* species are regularly parasitic on species whose pupae are concealed in the soil, etc.

Genus ITOPLECTIS Foerster

A fairly large, almost cosmopolitan genus with about 10 species in Europe. Six species are now known to occur in Britain. A seventh species, *I. curticauda* (Kriechbaumer) was once recorded incorrectly as British, as a result of a misidentification (see Stelfox, 1929 and Perkins, 1941). *Itoplectis* species are primary parasitoids of easily accessible but weakly concealed or cocooned, smallish lepidopterous pupae (exceptionally being able to attack pharate pupae, or even pharate adults, successfully) and regularly function also as facultative pseudohyperparasites, developing in ichneumonoid pupae (and possibly also prepupae) within their cocoons. This hyperparasitic tendency seems to be particularly developed in species such as *I. clavicornis*, in which it may even be obligatory. Sometimes other endopterygote groups in superficially similar cocoons are attacked.

Key to species

1 Gaster uniformly reddish orange, contrasting with black head, thorax and propodeum; tergites 2 and 3 puncto-reticulate. Hind tibia red. (Female fore tarsal claws simple. Fore wing length 4·5–9·2 mm. Ovipositor–hind tibia index about 0·8.) . . . .
. . . . **melanocephala** (Gravenhorst)
— Gaster mainly black, concolorous with head, thorax and propodeum, sometimes with sides and hind margins of gastral tergites reddish; tergites 2 and 3 punctate. Hind tibia black and white, at least in part. (Female fore tarsal claws often with a tooth-like lobe.) . . 2

2 Antenna subclavate, with distal segments (except the final one) transverse (Fig. 136). Epicnemium without a median depression behind fore coxae. Ovipositor–hind tibia index 0·4–0·5. (Fore wing length 5·4–6·0 mm. Hind tibia bicoloured, black with a white band. Upper hind corner of pronotum usually yellowish. Female fore tarsal claw with a tooth-like lobe.) . . . . **clavicornis** (Thomson)
— Antenna thickened, with distal segments elongate (Fig. 137) but occasionally 2 or 3 of them quadrate (especially in males). Epicnemium with a marked median depression behind fore coxae (usually difficult to see). Ovipositor–hind tibia index at least 0·7. 3

3 Hind tarsus uniformly blackish. . . . . **Pimpla turionellae** (see p. 81)
— Hind tarsus black and white banded. 4

4 Female. 5
— Male. 8

5 Fore tarsal claw with a very small tooth (Fig. 138) (sometimes entirely absent). (Fore wing length 4·5–5·2 mm. Hind tibia bicoloured, blackish with a white band. Gastral tergites black. Hind trochanter blackish.). . . . . **insignis** Perkins (female)
— Fore tarsal claw with a conspicuous tooth-like lobe (Fig. 139). 6

6 Pronotum with upper hind corner black; antenna with underside of pedicel black (rarely reddish distally); tegula sometimes also blackish but often pale (at least anteriorly). Hind tibia bicoloured, black with a white band (Fig. 140) (the distal black area sometimes brownish medially). Pterostigma usually blackish. (Fore wing length 3·7–7·1 mm. Gastral tergites black. Hind trochanter reddish.) . . . . . . **aterrima** Jussila (female)
— Pronotum with upper hind corner pale yellowish; antenna with under- and outer side of pedicel yellowish or reddish; tegula pale. Hind tibia tricoloured, black with white and red bands (Fig. 141) (the red area sometimes brownish dorsally). Pterostigma blackish to yellowish. 7

7 Hind trochanter mostly black. Hind tarsus with segment 5 shorter (Fig. 142). Gastral tergites black but often marked with red-brown laterally and posteriorly. Pterostigma usually blackish. (Fore wing length 3·4–7·7 mm.) . **maculator** (Fabricius) (female)
— Hind trochanter entirely red. Hind tarsus with segment 5 longer (Fig. 143). Gastral tergites black, usually only very narrowly marked on posterior margin with brownish-yellow. Pterostigma usually yellowish. (Fore wing length 2·7–8·5 mm.) . . . .
. . . . **alternans** (Gravenhorst) (female)

8 Mid and hind trochanters entirely yellow or red (rarely slightly brownish proximally). 9
— Mid and hind trochanters at least partly (hind usually mainly) blackish. 10

9 Pronotum with upper hind corner black; antenna with underside of pedicel black (rarely reddish distally); tegula sometimes also blackish but often pale (at least anteriorly). Hind tibia bicoloured, black with a white band (Fig. 140) (the distal black area sometimes brownish medially). Fore and mid coxae almost entirely black. . . . .
. . . . **aterrima** Jussila (male)
— Pronotum with upper hind corner pale yellowish; antenna with under- and outer side of pedicel yellowish or reddish; tegula pale. Hind tibia tricoloured, black with white and red bands (Fig. 141) (the red area sometimes brownish dorsally). Fore and mid coxae at least partly pale (sometimes entirely so). . . . . . . **alternans** (Gravenhorst) (male)

10 Antenna with underside of scape and pedicel entirely yellow. . . . .
. . . . **maculator** (Fabricius) (male)
— Antenna with underside of scape and pedicel blackish. . . . . **insignis** Perkins (male)

I. alternans. Widespread; common in the southern part of Britain, becoming rare northwards to Ross & Cromarty; also known from Ireland: Dublin. Flight period: iv–ix; bivoltine, overwintering as a prepupa. Occurs in a wide range of habitats but especially in open bushy places and tree canopy. Reared from many kinds of smallish Lepidoptera pupae, especially those cocooned or poorly concealed in more or less exposed or aerial vegetation, and as a facultative pseudohyperparasite from various ichneumonoid cocoons. Seen from pupae of the following Lepidoptera: *Taleporia tubulosa* (Retzius) (3 specimens), *Psyche casta* (Pallas) (1), *Sterrhopterix fusca* (Haworth) (1) (Psychidae), *Caloptilia cuculipennella* (Hübner) (1), *C. elongella* (Linnaeus) (1), *C. stigmatella* (Fabricius) (3), *Phyllonorycter lantanella* (Schrank) (1), *P. corylifoliella* (Hübner) (1), *P. ?quercifoliella* (Zeller) (2) (Gracillariidae), *Anthophila fabriciana* (Linnaeus) (2) (Choreutidae), *Glyphipterix haworthana* (Stephens) (1) (Glyphipterigidae), *Ypsolopha ustella* (Clerck) (1), *Yponomeuta padella* (Linnaeus) (1), *Yponomeuta* sp. (1) (Yponomeutidae), *Coleophora serratella* (Linnaeus) (3), *Coleophora saturatella* Stainton (2), *Coleophora* sp. (1) (Coleophoridae), *Carcina quercana* (Fabricius) (1), *Diurnea fagella* (Denis & Schiffermüller) (2) (Oecophoridae), *Anacampsis populella* (Clerck) (1) (Gelechiidae), *?Strophedra weirana* (Douglas) (1), *Ancylis unculana* (Haworth) (1), *Hedya nubiferana* (Haworth) (1), *Cacoecimorpha pronubana* (Hübner) (1), *Neosphaleroptera nubilana* (Hübner) (1), *Tortrix viridana* (Linnaeus) (1) (Tortricidae), *Drepana binaria* (Hufnagel) (1) (Drepanidae), *Thera juniperata* (Linnaeus) (3), and *?Thera* sp. (1) (Geometridae). Seen as a pseudohyperparasite via cocoons of *Apanteles* (sensu lato)/*Caloptilia* (2), *Aleiodes*/*Lomaspilis* (1) (Braconidae), *Hyposoter*/*Gonepteryx* (3), *Hyposoter*/*Orgyia* (2), *Casinaria*/indet. lepidopteran (1), *Phobocampe*/indet. lepidopteran (1), and indet. campoplegine(s) (2) (Ichneumonidae).

I. aterrima. In Britain this species has been mixed with *alternans* (partly as 'var. *kolthoffi*'; see Perkins, 1941). Kasparyan (1973) was the first to recognise it as a separate species. In view of its mainly boreal distribution in the Palaearctic it is surprising that Kasparyan did not investigate or comment on its possible relationship to *kolthoffi* (described from Greenland in 1890 by Aurivillius).

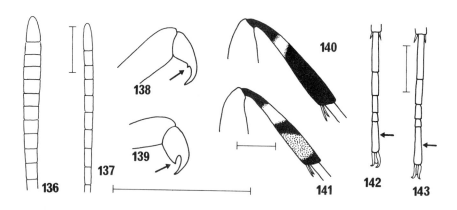

Figs 136–143. *Itoplectis* species. 136, apex of antenna, *I. clavicornis* female. 137, apex of antenna, *I. maculator* female. 138, fore tarsal claw, lateral view, *I. insignis* female. 139, fore tarsal claw, lateral view, *I. maculator* female. 140, left hind tibia, lateral view, *I. aterrima*. 141, left hind tibia, lateral view, *I. alternans*. 142, hind tarsus, dorsal view, *I. maculator* female. 143, hind tarsus, dorsal view, *I. alternans* female. All scale lines represent 0·5 mm.

Widespread in mainland Britain, but not so far found in Ireland. Uncommon in the south but becoming commoner northwards and especially so in the Scottish Highlands where it is often found above the tree line. Even in the south of Britain it has been collected chiefly in open places such as heaths, mosses and moors. Flight period: iv–ix, but certainly univoltine at high altitude. The host range seems very similar to that of *I. alternans*, given their habitat differences. Seen from pupae of the following Lepidoptera: *Sterrhopterix fusca* (Haworth) (1 specimen) (Psychidae), *Caloptilia betulicola* (Hering) (1) (Gracillariidae), and ?*Croesia bergmanniana* (Linnaeus) (2) (Tortricidae). Seen as a pseudohyperparasite via cocoons of *Aleiodes/Abraxas* (on *Calluna*) (1), *Aleiodes/Phragmatobia* (1), *Meteorus/Acleris* (1) (Braconidae), and *Casinaria/Zygaena* (3) (Ichneumonidae).

I. clavicornis. Rare (11 specimens from 7 localities); England: Devon, Hampshire, Surrey, Berkshire, Oxfordshire; Scotland: Argyll; [Ireland: Wicklow, Dublin (Stelfox, 1929), not examined], perhaps only in long established woodland. Flight period: v–vi; the winter is passed in the host cocoon according to Stelfox (1929). Associated with Lepidoptera mainly (or perhaps entirely) as a pseudohyperparasite, attacking cocoons of Ichneumonidae and Braconidae (see Stelfox, 1929; Townes, 1969; Aubert, 1969). We have seen two specimens reared on separate occasions from the 'jumping' cocoons of the ichneumonid *Scirtetes robustus* (Woldstedt) (= *Spudastica kriechbaumeri* (Bridgman)). The labelling on one indicated that it was itself one of two such rearings, and that the *Scirtetes* cocoons had been highly active after being collected from oak.

I. insignis. The British specimens that we have identified as this species differ slightly from Perkins' type series (which we have examined). One conspicuous difference is in the mid and hind trochanters and trochantelli which are more or less uniformly blackish, whereas in the types and other central European material the distal part of the trochanters and most of the trochantelli are whitish. The British material may represent an additional, undescribed species but no firm decision can be made on the basis of the few specimens available.

Rare (3 specimens from 3 localities); Scotland: Dunfermline, North East Fife, Badenoch & Strathspey. Flight period: v, viii. Perkins (1957) described this species as a parasitoid of the tortricid moth *Zeiraphera diniana* (Guenée) (as *Eucosma griseana* (Hübner)) in Switzerland, but from only two specimens.

I. maculator. Very common, distributed throughout Great Britain and Ireland. Flight period: iii–vii, ix–x; univoltine with a single emergence about vi–vii when the sexes mate, the females subsequently aestivating (e.g. in grass tussocks) through viii and reappearing to feed on honeydew etc. in ix–x before overwintering as adults (e.g. in conifer foliage), resuming feeding in spring and attacking their hosts in about vi (Cole, 1967). The very few reared specimens dated viii and ix are possibly indicative of a small second generation. Although it occurs in a wide variety of habitats, and encompasses a broad host range in so doing, *I. maculator* has a life cycle particularly adapted to exploiting the high density spring flush of tortricine Tortricidae, many of which are arboreal and pupate around June in folded leaves on trees or in the understorey of deciduous woodland, where *I. maculator* is correspondingly common. Seen from pupae (occasionally pharate) of the following Lepidoptera: *Acanthopsyche atra* (Linnaeus) (1 specimen), *Sterrhopterix fusca* (Haworth) (1) (Psychidae), *Aspilapteryx tringipennella* (Zeller) (2) (Gracillariidae), *Prochoreutis myllerana* (Fabricius) (1), *Prochoreutis* sp. (1) (Choreutidae), *Coleophora serratella* (Linnaeus) (4), *Coleophora* spp. (2) (Coleophoridae), *Agonopterix assimilella* (Treitschke) (2) (Oecophoridae), *Anacampsis populella* (Clerck) (3) (Gelechiidae), *Mompha conturbatella* (Hübner) (1) (Momphidae), *Gypsonoma dealbana* (Frölich) (1), *Lobesia occidentis* Falkovitsh (2), *Pandemis cerasana* (Hübner) (2), *Archips crataegana* (Hübner) (2), *A. rosana* (Linnaeus) (6), *Clepsis spectrana* (Treitschke) (2), *Ptycholoma lecheana* (Linnaeus) (5), *Ditula angustiorana* (Haworth) (1), *Cnephasia longana* (Haworth) (1), *C. stephensiana* (Doubleday) (1), *Tortrix viridana* (Linnaeus) (46; includes 27 from one concerted survey of this host), *Croesia bergmanniana* (Linnaeus) (1), *Acleris comariana* (Lenig & Zeller) (1), ?*A. hastiana* (Linnaeus) (1) (Tortricidae), and *Acrobasis consociella* (Hübner) (1) (Pyralidae). Seen as a pseudohyperparasite via cocoons of *Apanteles* (sensu lato)/*Cnephasia* (1 specimen), *Lissogaster* (= *Microgaster*)/*Aspilapteryx* (1), *Lissogaster* (= *Microgaster*)/*Vanessa* (2), *Dinocampus* (= *Perilitus*)/*Coccinella* (1), ?*Charmon*/indet. lepidopteran (1), *Aleiodes*/indet. noctuid (2), *Aleiodes*/*Philudoria* (1), *Aleiodes*/*Thymelicus* (1), *Aleiodes*/*Leucoma* (6) (Braconidae), *Casinaria*/*Zygaena* (14), *Phobocampe*/*Operophtera* (1), *Phobocampe*/indet. lepidopteran (2), *Hyposoter*/ *Abraxas* (1), *Hyposoter*/*Gonepteryx* (1), indet. campolegine (3), *Glypta*/*Olethreutes* (1), *Glypta*/indet. tortricid (5), and *Phytodietus*/indet. lepidopteran (1) (Ichneumonidae). Also seen reared from cocoons of *Hypera* spp. (3) (Coleoptera: Curculionidae).

I. melanocephala. Rare, strongly associated with *Phragmites* and probably restricted to old fens; England: Devon, Dorset, Berkshire, Leicestershire, Cambridgeshire, Suffolk, Norfolk. Flight period: v–viii. We have seen small individuals reared from pupae of the oecophorid moths *Depressaria pastinacella* (Duponchel) (1 specimen) and *D. daucella* (Denis & Schiffermüller) (1), both of which pupate inside the hollow stems of .Umbelliferae. The majority of caught specimens seen are too big to have used these hosts, however, and it seems likely that it attacks the rather larger Lepidoptera pupating in *Phragmites* stems more regularly.

Genus **PIMPLA** Fabricius

Pimpla is a large cosmopolitan genus with about 15 species in Europe. The larger species such as *P. hypochondriaca* are among the relatively few ichneumonids that attract the attention of the general public. It is particularly unfortunate, therefore, that the name of the genus should be disputed: many workers (but principally Townes) refer to it by the junior synonym *Coccygomimus*, because they believe the name *Pimpla* should be applied to the genus called *Ephialtes* in this handbook. This nomenclatural tangle, which also involves the pimpline genus *Apechthis*, was discussed by Townes (1969), Fitton & Gauld (1976) and Carlson (1979). It has not yet been satisfactorily resolved.

Unlike *Itoplectis* and *Apechthis* species, several species of *Pimpla* undoubtedly parasitise lepidopterous pupae concealed in moss, soil, etc. and accordingly families such as Geometridae and Noctuidae are probably more extensively involved in host ranges. Other endopterygotes with fairly hard cocoons or pupal cuticles enter host ranges occasionally, and pseudohyperparasitism is also known. Except as noted, the bodies of all British species are black or blackish.

Key to species

1	Female	2
—	Male	11

2   Hair on face black or dark brown (this character sometimes causes difficulty; care must be taken to distinguish between the colour of the hairs and the bright reflections from their surface). (Hair on propodeum usually blackish, if pale then malar space greater than basal width of mandible.) **3**

—   Hair on face pale, usually whitish but sometimes pale yellowish. (Hair on propodeum pale and malar space often not greater than basal width of mandible.) **6**

3   Mesopleuron centrally closely punctate, the distance between the punctures less than or about equal to their diameters. **4**

—   Mesopleuron centrally more sparsely punctate, the punctures separated by about twice their diameters. **5**

4   Upper surface of scutellum shining and sparsely, but usually coarsely, punctate. Legs reddish with only coxae, trochanters, trochantelli and hind tarsi blackish. (Fore wing length 4·8–14·9 mm. Ovipositor–hind tibia index 0·8–1·0.) . **hypochondriaca** (Retzius) (female)

—   Upper surface of scutellum matt, closely punctate. Legs predominantly blackish, hind ones entirely so. (Fore wing length 10·2–13·4 mm. Ovipositor–hind tibia index 0·6–0·7.) . . . . **aethiops** Curtis (female)

5   Malar space at most equal to basal width of mandible. Head in dorsal view evenly constricted behind the eyes (Fig. 151). Hind coxa red. (Fore wing length 6·8–13·7 mm. Ovipositor–hind tibia index 0·9–1·0.) . . . . . . **arctica** Zetterstedt (female)

—   Malar space conspicuously greater than basal width of mandible. Head in dorsal view not as constricted behind the eyes (Fig. 152). Hind coxa black. (Fore wing length 4·0–8·4 mm. Ovipositor–hind tibia index 0·9–1·0.) . . . . **sodalis** Ruthe (female)

6   Mesopleuron strongly and closely punctate, the punctures separated by about their own diameters. Pronotum usually with a yellow stripe on hind corner. Interocellar distance about 2·0 times orbital-ocellar distance (Fig. 153). (Fore wing length 3·0–11·6 mm. Ovipositor–hind tibia index 0·9–1·1. Hind coxa black.). . . . . **turionellae** (Linnaeus) (female)

—   Mesopleuron finely punctate, the punctures separated by about two or more times their diameters. Pronotum without a yellow stripe. Interocellar distance at most about 1·5 times orbital–ocellar distance (Fig. 154). **7**

7   Laterotergite of tergite 4 of gaster relatively narrow, about twice as long as broad; laterotergite of tergite 3 not much wider posteriorly than anteriorly (Fig. 144). Tergites of gaster weakly to very obviously coriaceous between punctures. Propodeum with or without carinae delineating an area basalis-superomedia.. . . . . **8**

—   Laterotergite of tergite 4 of gaster wide, almost as long as broad; laterotergite of tergite 3 rather wider posteriorly than anteriorly (Figs 145, 146). Tergites of gaster very weakly coriaceous or smooth between punctures. Propodeum with at least weak carinae delineating an area basalis-superomedia. **9**

8   Propodeum without carinae defining a combined area basalis-superomedia. Hind tibia blackish with a well defined whitish band near proximal end. Underside of proximal part of antennal flagellum yellowish. Mid coxa red. (Fore wing length 3·8–8·5 mm. Ovipositor–hind tibia index 0·6–0·8. Scutellum sometimes yellow marked. Hind coxa and femur red.) . . . . **flavicoxis** Thomson (female)

—   Propodeum with at least weak carinae delineating a combined area basalis-superomedia. Hind tibia blackish with a poorly defined dark reddish band near proximal end. Underside of proximal part of antennal flagellum blackish. Mid coxa black. (Fore wing length 4·5–6·8 mm. Ovipositor–hind tibia index 0·9–1·1. Hind coxa and femur red.) . . . . **wilchristi** sp. nov. (female)

9   Laterotergite of tergite 2 of gaster narrower than that of tergite 4, as in Fig. 145. Hind femur black-marked at distal apex. Face with rather fine punctures. (Fore wing length 3·7–7·8 mm. Ovipositor–hind tibia index about 1·0. Hind coxa red, hind tibia infuscate, white banded.) . . . . **melanacrias** Perkins (female)

—   Laterotergite of tergite 2 of gaster almost as wide as that of tergite 4 (Fig. 146). Hind femur entirely red. Face with rather fine to moderate punctures **10**

10   Ovipositor–hind tibia index 0·9–1·1. Fore tarsus with segment 4 dorsally about as long as broad (Fig. 155). (Fore wing length 3·3–7·6 mm. Hind coxa red, hind tibia infuscate with a yellowish or reddish band.). . . . . **spuria** Gravenhorst (female)

—   Ovipositor–hind tibia index 0·6–0·8. Fore tarsus with segment 4 dorsally only about 0·5 as long as broad (Fig. 156). (Fore wing length 3·3–9·0 mm. Hind coxa red, hind tibia infuscate with a white band.) . . . . **contemplator** (Müller) (female)

11 Hair on face blackish. (Laterotergite of tergite 3 of gaster about 3 times as long as broad.)
    12
—   Hair on face whitish. 15
12 Antennal flagellum with segments 6–9 (at least) each with a tyloid (a longitudinal, raised,
    polished area) on its outer side. 13
—   Segments of antennal flagellum without tyloids.. 14
13 All legs with femur and tibia red.. . . . . **hypochondriaca** (Retzius) (male)
—   Fore and mid legs with femur blackish and yellow and tibia yellowish, hind leg with femur
    and tibia entirely blackish. . . . . **aethiops** Curtis (male)
14 Head in dorsal view evenly constricted behind eyes (Fig. 151). Tergite 1 of gaster in profile
    distinctly angled behind level of spiracles (Fig. 157). Hind coxa usually partially red.
    . . . . **arctica** Zetterstedt (male)
—   Head in dorsal view not narrowed behind eyes (Fig. 152). Tergite 1 of gaster in profile not
    distinctly angled behind spiracle (Fig. 158). Hind coxa black.. **sodalis** Ruthe (male)
15 Interocellar distance about twice orbital-ocellar distance (Fig. 153). Mesoscutum strongly
    and closely punctate.. . . . . **turionellae** (Linnaeus) (male)
—   Interocellar distance at most 1·5 times orbital-ocellar distance (Fig. 154). Mesoscutum with
    fine shallow punctures. 16
16 Fore and mid legs. Laterotergite of tergite 5 of gaster about 3 times as long as broad
    (Fig. 147) AND hind tibia blackish with a well defined whitish band near proximal end.
    (Tergites 2–3 of gaster with area between punctures matt, finely coriaceous. Scutellum
    sometimes yellow marked.). . . . . **flavicoxis** Thomson (male)
—   Fore and mid coxae black. Laterotergite of tergite 5 of gaster 2 to 3 times as long as broad
    (Figs 148–150), if about 3 times (Fig. 148) then hind tibia blackish with a poorly defined
    reddish band near proximal end (otherwise with a whitish band). 17
17 Laterotergite of tergite 2 of gaster narrow, at least 5 times as long as broad (Figs 148, 149).
    18
—   Laterotergite of tergite 2 of gaster 2·5 or less times long as broad (Fig. 150). (Hind femur red,
    rarely slightly infuscate at distal apex.) 19
18 Hind femur entirely red, rarely slightly infuscate at distal apex. Hind tibia blackish with a
    reddish band near proximal end. (Laterotergite of tergite 5 of gaster about 3 times as long
    as broad (Fig. 148). Tergites 2–3 of gaster with area between punctures matt, finely
    coriaceous.). . . . . **wilchristi** sp. nov. (male)
—   Hind femur red, rather sharply black marked at distal apex. Hind tibia blackish with a
    whitish band near proximal end. (Laterotergite of tergite 5 of gaster about 2 times as
    long as broad (Fig. 149). Tergites 2–3 with area between punctures shining, smooth.)
    . . . . **melanacrias** Perkins (male)
19 Malar space about equal to basal width of mandible. Tergite 6 of gaster clearly punctate in
    anterior half. . . . . **spuria** Gravenhorst (male)
—   Malar space 0·7–0·8 of basal width of mandible. Tergite 6 of gaster at most distinctly
    punctate at only extreme anterior edge.. . . . . . **contemplator** (Müller) (male)

P. aethiops. Probably extinct; 12 specimens examined, but only one with locality data:
England: Greater Manchester, Bowden, [no date]; but it has also been recorded (as
aterrima Gravenhorst) from Netley in Shropshire (Gravenhorst, 1829). Except for the
Bowden specimen, which was probably collected in the period 1870–1920 but for which
the locality data may be unreliable (C. Johnson, pers. comm.), the British records all
pre-date 1868. Flight period: unknown. Some of the British material was reared from
Papilio machaon Linnaeus (Papilionidae) (1 specimen) and *Laelia coenosa* (Hübner)
(Lymantriidae) (4), indicating that it once occurred in the fenlands of East Anglia.

P. arctica. Very rare (6 specimens from 1 locality); Scotland: Skye & Lochalsh, Isle of
Soay, ix.1909. This is a boreo-alpine species, of which we have seen one Swiss specimen
reared from an unidentified psychid case.

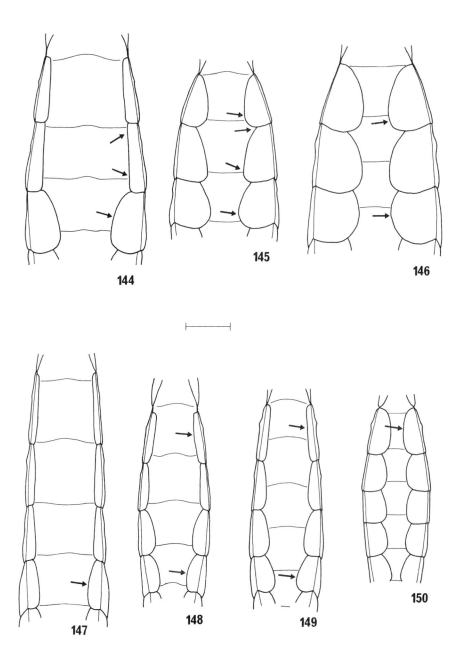

Figs 144–150. *Pimpla* species, gaster, ventral view. 144, segments 2–4, *P. wilchristi* female. 145, segments 2–4, *P. melanacrias* female. 146, segments 2–4, *P. spuria* female. 147, segments 2–5, *P. flavicoxis* male. 148, segments 2–5, *P. wilchristi* male. 149, segments 2–5, *P. melanacrias* male. 150, segments 2–5, *P. spuria* male. Scale line represents 0·5 mm.

P. contemplator. Common; widely distributed in wooded areas in the south of Britain north to Gwynedd and Cumbria. Flight period: v–x; bivoltine, overwintering as a prepupa. It attacks smallish lepidopterous pupae found shallowly concealed in the soil, leaf litter etc, and we have seen it from *Tortricodes alternella* (Denis & Schiffermüller) (1 specimen), *Zeiraphera insertana* (Fabricius) (2) (Tortricidae), and *Operophtera brumata* (Linnaeus) (3) (Geometridae), as well as numerous examples reared in one survey from the soil and leaf litter below oak trees. We have also seen one specimen reared from the puparium of the hoverfly *Dasysyrphus tricinctus* (Fallén), similarly collected from leaf litter.

P. flavicoxis. This has been considered synonymous with the North American species *aquilonia* (for example, by Townes & Townes, 1960; Townes *et al.*, 1965; Oehlke, 1967; and Aubert, 1969). More recently (Kasparyan, 1974) it has again been recognised as a separate species. Its relationship with *aquilonia* awaits further investigation, preferably in the context of a comprehensive revision of the Holarctic *Pimpla* species.

Very common throughout the British Isles, occurring in a wide range of habitats. Flight period: v–ix; bivoltine, at least in the south, overwintering as a prepupa. It commonly attacks smallish lepidopterous pupae poorly concealed on trees and bushes, but it has also been reared from those in moss and from the soil below oak trees. We have seen specimens from the pupae of *Carcina quercana* (Fabricius) (1 specimen) (Oecophoridae), *Epiblema roborana* (Denis & Schiffermüller) (1), *Eudemis profundana* (Denis & Schiffermüller) (2), *Ditula angustiorana* (Haworth) (1), *Tortrix viridana* (Linnaeus) (2) (Tortricidae), *Thera juniperata* (Linnaeus) (5), and *Operophtera*

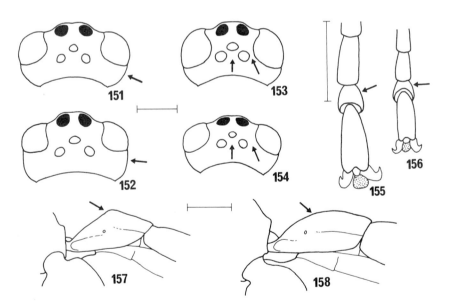

Figs 151–158. *Pimpla* species. 151, head, dorsal view, *P. arctica* male. 152, head, dorsal view, *P. sodalis* male. 153, head, dorsal view, *P. turionellae* male. 154, head, dorsal view, *P. wilchristi* male. 155, segments 3–5 of fore tarsus, dorsal view, *P. spuria* female. 156, segments 3–5 of fore tarsus, dorsal view, *P. contemplator* female. 157, segment 1 of gaster, lateral view, *P. arctica* male. 158, segment 1 of gaster, lateral view, *P. sodalis* male. All scale lines represent 0·5 mm.

brumata (Linnaeus) (1) (Geometridae). We have also seen one specimen reared as a pseudohyperparasite from ?*Campoletis*/?*Diachrysia* (Ichneumonidae).

P. hypochondriaca. It is particularly unfortunate that we have had to change the name of this species. Previously it was known as *Pimpla instigator* (Fabricius) (or sometimes *Coccygomimus instigator*) and as such was one of the handful of ichneumonid names which meant something to non-ichneumonid workers. *Ichneumon instigator* Fabricius, 1793 is a junior primary homonym (of *Ichneumon instigator* Rossius, 1790; which appears to be a pimpline but not a species of *Pimpla*) and must be permanently rejected under the *International Code of Zoological Nomenclature*. This fact must surely have been known to several earlier workers but presumably they conveniently overlooked it. Our search for the earliest available name for this species was driven by the expectation that such a common, widespread, large and conspicuous insect would have been described before 1793, and indeed both Dalla Torre (1901) and Oehlke (1967) listed several potential senior synonyms. These were not even mentioned by Perkins (1941) or Aubert (1969). The situation was not helped by the International Commission on Zoological Nomenclature when they designated (in Opinion 159, 1945) *Ichneumon instigator* Fabricius as the type species of *Pimpla* without checking on something as simple as primary homonymy (see Sherborn, 1902).

The species is here identified as *I. hypochondriacus* Retzius and the detailed formal synonymy is given below.

Pimpla hypochondriaca (Retzius)

Ichneumon hypochondriacus Retzius, 1783: 67. Holotype female, SWEDEN (Degeer collection, Naturhistoriska Riksmuseet, Stockholm) (lost) [missing from the collection as early as 1810, according to a manuscript catalogue (T. Kronestedt, pers. comm.)].
Ichneumon inguinalis Geoffroy in Fourcroy, 1785: 397. Type(s) [?sex], FRANCE (lost). **syn. nov.**
Ichneumon instigator Fabricius, 1793: 164. Lectotype female, GERMANY (Zoologisk Museum, Copenhagen), by designation of Townes, Momoi & Townes, 1965: 51 [not examined]. **syn. nov.**

Retzius's (1783) publication served primarily to give a formal name to the species described and figured by Degeer (1771: 845–847 and plate 29, figures 6–8). It is unfortunate that the characters which Retzius chose to include in his brief diagnosis seem at first sight to be inapplicable to this species. However, Degeer's description, figures and account of rearing of his ichneumonid leave no room for doubt about the identity of the species, even though the specimen is no longer extant.

It is impossible to be certain that *I. inguinalis* Geoffroy is this species but, in the absence of specimens and because all of the details given in the brief description fit, it is here formally synonymised with *hypochondriaca*. The 'point blanc à la base des cuisses' mentioned by Geoffroy is undoubtedly the dorsal membrane between the coxa and the trochanter, which is pale in colour and relatively conspicuous in specimens with the femora extended.

We have not reconsidered the identity of the nominal species post-dating 1793 which have been synonymised with *instigator* (see Perkins, 1941; Oehlke, 1967; Aubert, 1969; Kasparyan, 1974). The other potential senior synonyms of *instigator* listed by Dalla Torre and/or Oehlke are all rejected as not applying to this species, for the reasons given below.

Ichneumon compunctator Poda, 1761: 106. This is an emendation of *I. compunctor* Linnaeus (see Poda, 1761: 105, where he states 'Fortassis Compunctator Linn. antennae deficiunt.'). *Ichneumon compunctor* Linnaeus is a species of *Apechthis*.

I. punctator Allioni, 1766: 196. (Dalla Torre gave the author incorrectly as Müller.) This species has not been identified but the colour pattern of the abdomen ('Abdomen subtus albidum utrinque quator nigris') indicates that it cannot be *hypochondriaca*.

I. punctator: Müller, 1776: 157. This is a reference to *I. punctator* Linnaeus and not the description of a new species. The Linnaean species belongs to *Xanthopimpla* (Townes & Chui, 1970: 39).

I. bisulcus Gmelin, 1790: 2698. The type material is lost. The notauli are not conspicuous in *hypochondriaca* so it seems unlikely that it is this species.

I. impressus Gmelin, 1790: 2698. The type material is lost. This species is described as having the vertex impressed, which is not the case in *hypochondriaca*.

I. fornicator: Rossius, 1790: 44. This is a reference to *I. fornicator* Fabricius and not the description of a new species. The Fabrician species belongs to the genus *Exetastes* (Townes *et al.*, 1965: 229).

I. faber Christ, 1791: 364. The hind femora are described and illustrated as being black, which is never the case in *hypochondriaca*. However, the description has some peculiar features and there are discrepancies between it and the figure.

P. hypochondriaca is very common, especially in hedgerows, gardens and similar situations, and widely distributed in the British Isles north to Ross & Cromarty. Flight period: vi–x; bivoltine, overwintering as a prepupa. It attacks the naked or cocooned pupae of medium-sized to large Lepidoptera that pupate above ground, particularly on tall vegetation, bushes, fences etc. It frequently invades or emerges in lepidopterists' rearing cages, and this may be how it has gained access to the few hosts recorded below that normally pupate below ground level or in some other deeply concealed site. We have seen it reared from the pupae of *Zygaena* sp. (1 specimen) (Zygaenidae), ?*Olethreutes lacunana* (Denis & Schiffermüller) (1) (Tortricidae), *Papilio machaon* Linnaeus (3) (Papilionidae), *Pieris brassicae* (Linnaeus) (26), *Pieris rapae* (Linnaeus) (3), *Pieris* sp. (2) (Pieridae), *Lycaena dispar batavus* (Oberthür) (20, from one concerted survey of this host) (Lycaenidae), *Eriogaster lanestris* (Linnaeus) (1), *Malacosoma neustria* (Linnaeus) (6), *Lasiocampa quercus* (Linnaeus) (1), *Philudoria potatoria* (Linnaeus) (1) (Lasiocampidae), *Saturnia pavonia* (Linnaeus) (1) (Saturniidae), *Drepana binaria* (Hufnagel) (2) (Drepanidae), *Tethea ocularis* (Linnaeus) (1) (Thyatiridae), *Abraxas grossulariata* (Linnaeus) (1), *Ennomos autumnaria* (Werneburg) (1), *Biston betularia* (Linnaeus) (1), *Peribatodes rhomboidaria* (Denis & Schiffermüller) (2), *Alcis jubata* (Thunberg) (1) (Geometridae), *Mimas tiliae* (Linnaeus) (1), *Deilephila porcellus* (Linnaeus) (1) (Sphingidae), *Cerura vinula* (Linnaeus) (1; failed to escape from the hard host cocoon) (Notodontidae), *Orgyia recens* (Hübner) (3), *O. antiqua* (Linnaeus) (11), *Euproctis chrysorrhoea* (Linnaeus) (7), *Leucoma salicis* (Linnaeus) (2) (Lymantriidae), *Lithosia quadra* (Linnaeus) (3) (Arctiidae), *Ceramica pisi* (Linnaeus) (1), *Cucullia verbasci* (Linnaeus) (1), *Aporophyla australis* (Boisduval) (1), *Moma alpium* (Osbeck) (1), *Acronicta euphorbiae* (Denis & Schiffermüller) (2), *A. rumicis* (Linnaeus) (2), *Acronicta* sp. (1), *Gortyna flavago* (Denis & Schiffermüller) (1), and *Calocasia coryli* (Linnaeus) (1) (Noctuidae). We have also seen two specimens, with host remains, reared from *Rhagium* sp. (Coleoptera: Cerambycidae); and three specimens labelled as from *Trichiosoma* sp. (Hymenoptera: Cimbicidae), one of which was, however, accompanied by a lasiocampid cocoon (?*Trichiura crataegi* (Linnaeus)). The willingness of *P. hypochondriaca* to exploit a broad host range is amusingly illustrated by Halstead's (1987) account of a specimen attempting to oviposit into bean seeds detected as hard swellings within a packet, no doubt perceived as cocooned pupae.

P. melanacrias. Uncommon in southern England but becoming commoner in Scotland, especially on high moorland, and also in Ireland but not so far found in Wales. Flight period: vi–ix. There are no host records.

P. sodalis. Rare (13 specimens from one area); Scotland: Badenoch & Strathspey. A boreoalpine species with a circumpolar distribution; in Britain known only from the north-west side of the Cairngorm plateau at about 800–1200 m (Perkins (1941) records some of this material as coming from Perthshire in error). Flight period: v–vii. Nuorteva & Jussila (1967; 1969) and Jussila & Nuorteva (1968) presented circumstantial evidence that this species may opportunistically attack the geometrid moth *Epirrita autumnata* (Borkhausen) in the Fennoscandian mountain birch zone, but in Scotland it occurs well above the tree line and perhaps exculsively so. We have seen no reared material.

P. spuria. Uncommon but widely distributed in the British Isles north to Glasgow City and North East Fife. Flight period: v–ix; bivoltine, overwintering as a prepupa. Commonly reared from pupae of depressariine oecophorid moths in the hollow stems of Umbelliferae, but it clearly attacks a wider range of smallish Lepidoptera in tall field layer vegetation. We have seen it reared from *Tebenna bjerkandrella* (Thunberg) sensu British authors (2 specimens) (Choreutidae), *Depressaria pastinacella* (Duponchel) (16) (Oecophoridae), *Cynaeda dentalis* (Denis & Schiffermüller) (1), *Oncocera genistella* (Duponchel) (1) (Pyralidae), and (? in captivity) *Eupithecia vulgata* (Haworth) (3) (Geometridae).

P. turionellae. Common, particularly where there are trees, and widespread in the British Isles. Flight period: v–x; bivoltine, perhaps sometimes overwintering as an adult. It attacks cocooned, or more rarely naked, smallish to medium-sized pupae, including those well concealed in substrates such as tree bark. Most of the available reared material is from hosts that feed and pupate on trees and bushes including conifers. We have seen it reared from pupae of *Taleporia tubulosa* (Retzius) (1 specimen), *Luffia ferchaultella* (Stephens) (1), *Psyche casta* (Pallas) (1) (Psychidae), *Yponomeuta evonymella* (Linnaeus) (10), *Yponomeuta cagnagella* (Hübner) (1), *Yponomeuta* sp. (3), *Ypsolopha vittella* (Linnaeus) (2) (Yponomeutidae), *Depressaria pastinacella* (Duponchel) (1) (Oecophoridae), *Limnaecia phragmitella* Stainton (6) (Momphidae), *Cydia pomonella* (Linnaeus) (1), *Rhyacionia buoliana* (Denis & Schiffermüller) (30; includes 27 from one concerted survey of this host), *Zeiraphera diniana* (Guenée) (1), *Olethreutes bifasciana* (Haworth) (1), *Tortrix viridana* (Linnaeus) (4) (Tortricidae), *Acrobasis consociella* (Hübner) (2), *Oncocera genistella* (Duponchel) (5), *Metriostola betulae* (Goeze) (1) (Pyralidae), *Pieris rapae* (Linnaeus) (2) (Pieridae), *Strymonidia w-album* (Knoch) (2) (Lycaenidae), *Abraxas grossulariata* (Linnaeus) (4), *Alcis jubata* (Thunberg) (1) (Geometridae), *Orgyia antiqua* (Linnaeus) (3), and *Euproctis similis* (Fuessly) (2) (Lymantriidae). We have also seen one specimen reared as a pseudohyperparasite from *Meteorus/Acrobasis* (Braconidae), and various workers (e.g. Sandlan, 1980) have maintained laboratory cultures on tenebrionid beetle pupae.

P. wilchristi sp. nov.

Figs 44, 47, 52, 144, 148, 154, 159.

Female: Fore wing length 4·5–6·8 mm. Hair on face pale, yellowish (in one specimen with a few darker hairs intermixed). Malar space a little less than basal width of mandible. Upper surface of scutellum shining, with moderately fine punctures. Mesopleuron finely punctate. Lower posterior part of the mesopleuron and the metapleuron with moderately strong oblique striae. Tergites of gaster weakly coriaceous between the strong, dense punctures; laterotergites as in Fig. 144. Ovipositor–hind tibia index 0·9–1·1.

Colour mainly black; with the following red or reddish: ovipositor, all femora, fore and mid tibiae and tarsi, mid and hind trochantelli, hind coxa, and an ill-defined band on the hind tibia. Other parts, such as the posterior margins of the gastral tergites and the distal section of the hind trochanter, sometimes reddish. Wings weakly infuscate; with the veins black, except most of fore wing costa and the proximal part of the pterostigma, which are pale.

Male: Similar to female except laterotergites of gaster as in Fig. 148 and all coxae black.

This species resembles *P. flavicoxis* but differs as noted in couplets 8 and 16/18 in the key, and in having the laterotergites of the gaster slightly wider; the hairs on the face pale yellowish; and the lower posterior part of the mesopleuron and the metapleuron with stronger oblique striae. The latter character also serves most easily to distinguish it from *P. arctica*. The male might be confused with *P. melanacrias*, which differs from it in the head less transverse in front view, temples longer and less narrowed behind the eyes, and ocelli smaller, in addition to the characters in the key.

HOLOTYPE female: WALES: Gwynedd, Anglesey, Llangristiolus (Grid ref. SH434736), Malaise trap by hayfield, 7–27.viii.1982 (*Wilkinson*) (NMS).

PARATYPES, 14 females and 3 males, same data as holotype except dates (range 16.vii–25.ix.1982) (NMS and BMNH); 1 female, SCOTLAND: Cunninghame, Arran, Blackwater Foot, viii.1921 (*Campbell*) (NMS); 1 female, SCOTLAND: Kyle &

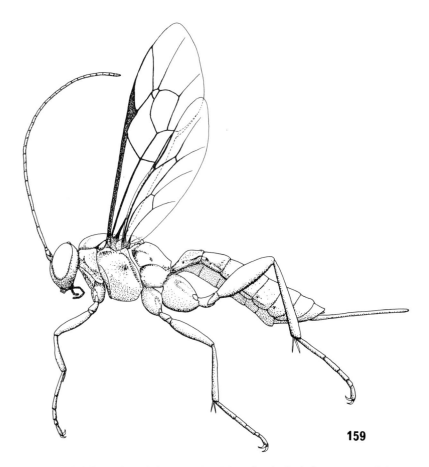

159

Fig. 159. Whole insect, lateral view, *Pimpla wilchristi* female. Scale line represents 0·5 mm.

Carrick, Ailsa Craig, The Trammins, 16.vi.1984 (*Christie*) (NMS); 1 female, SCOTLAND: North East Fife, Tentsmuir, 16.vii.1977 (*Noyes, Rogers & Huddleston*) (BMNH).

Three of the sites at which it has been collected are coastal, and the fourth (Llangristiolus) is less than 10 km inland: all are rather open and exposed. There are no rearing records.

Genus APECHTHIS Foerster

A moderately large genus with species in most regions of the world. Four species are known from Europe, of which three occur in Britain. Females of *Apechthis* are very easily recognised by the characteristic decurved apex of the ovipositor. Experiments by Cole (1959) illustrate the sucess of this structure as a specialisation for insertion between the overlapping plates or rings of hardened cuticle of somewhat wriggly, exposed or poorly concealed pupae of Lepidoptera.

Some workers, notably Townes, call this genus *Ephialtes*, with authorship attributed to Schrank, 1802. However, in Opinion 159 (1945), the International Commission on Zoological Nomenclature ruled that *Ephialtes* should be used in the sense of Gravenhorst (1829) and authorship attributed to him.

Key to species

1 Female. 2
— Male. 4
2 Hind tarsal claws simple, without a tooth-like basal lobe (see Fig. 57). (Fore wing length 5·1–11·6 mm. Ovipositor–hind tibia index 0·5–0·7. Black species with pronotum at hind corner, orbits, scutellum and postscutellum usually pale yellow marked; legs mainly red, hind tibia infuscate with a white sub-proximal band, hind tarsus infuscate.)
. . . . **quadridentata** (Thomson) (female)
— Hind tarsal claws with a tooth-like basal lobe (sometimes rather small, particularly on the inner claw) (see Fig. 56). 3
3 Hair on propodeum mainly brownish. Hind tibia mainly red, sometimes slightly infuscate at ends, never with a sub-proximal white band. (Fore wing length 7·3–12·0 mm. Ovipositor–hind tibia index 0·7–0·8. Coloration otherwise similar to *quadridentata*.)
. . . . **compunctor** (Linnaeus) (female)
— Hair on propodeum whitish. Hind tibia usually infuscate and with a whitish band near proximal end. (Fore wing length 5·4–12·7 mm. Ovipositor–hind tibia index 0·7–0·8. Coloration otherwise similar to *quadridentata* but usually with two yellow stripes on mesoscutum.) . . . . **rufata** (Gmelin) (female)
4 Hair on propodeum brown or black. Hind tibia entirely red, rarely darkened at ends. Face yellow, usually with a central black area (variable in extent). . . . .
. . . . **compunctor** (Linnaeus) (male)
— Hair on propodeum white. Hind tibia with a sub-proximal white band. Face usually entirely yellow. 5
5 Tergite 6 of gaster more evenly, coarsely punctured (anteriorly punctures separated by much less than their diameters, posteriorly by about their diameters), the interspaces finely alutaceous. (Posterior part of mesoscutum usually with a pair of yellow stripes.)
. . . . **rufata** (Gmelin) (male)
— Tergite 6 of gaster coarsely punctured anteriorly (punctures separated by less than their diameters), becoming finer and weaker posteriorly (punctures separated by at least twice their diameters), the interspaces shining. (Posterior part of mesocutum always entirely black.) . . . . **quadridentata** (Thomson) (male)

A. compunctor. Uncommon, occuring in more open situations that the other two species; widely distributed in the southern half of Britain: Dorset to Essex and as far north as Clwyd, Lancashire and Humberside. Flight period: v–vi, viii–x; bivoltine, overwintering as a prepupa. We have seen reared material from the pupae of the following Lepidoptera: *Ptycholoma lecheana* (Linnaeus) (1 specimen) (Tortricidae), *Pleuroptya ruralis* (Scopoli) (1) (Pyralidae), *Pieris brassicae* (Linnaeus) (1) (Pieridae), *Strymonidia pruni* (Linnaeus) (2), *Lycaena dispar batavus* (Oberthür) (1) (Lycaenidae), *Scoliopteryx libatrix* (Linnaeus) (1) (Noctuidae). Exposed butterfly pupae are clearly an important part of its host range (see also Aubert, 1969, and references therein).

A. quadridentata. Locally common in deciduous woodland and parks; widely distributed in Britain north to Argyll & Bute, and also in Ireland. Flight period: v–x; bivoltine, overwintering as a prepupa. Cole (1967) discussed the seasonal activity, seasonal dimorphism and alternation of hosts of this species (as *resinator* (Thunberg)) in detail. We have seen specimens (including Cole's) reared from the pupae of the following Lepidoptera: *Ptycholoma lecheana* (Linnaeus) (2 specimens), *Aleimma loeflingiana* (Linnaeus) (1), *Tortrix viridana* (Linnaeus) (28; includes 23 from one concerted survey of this host) (Tortricidae), *Lycaena dispar batavus* (Oberthür) (1) (Lycaenidae), and *Pararge aegeria* (Linnaeus) (4) (Satyridae).

A. rufata. Locally common in deciduous woodland and parks; widely distributed in Britain north to Stirling; and also found in Ireland: Down. Flight period: v–xi; bivoltine, overwintering as a prepupa. Cole (1967) recorded the seasonal activity, seasonal dimorphism and host alternation in detail. We have seen specimens (including Cole's) reared from pupae of the following Lepidoptera: *Carcina quercana* (Fabricius) (2 specimens), *Diurnea fagella* (Denis & Schiffermüller) (4) (Oecophoridae), *Pandemis cerasana* (Hübner) (1), *Archips xylosteana* (Linnaeus) (3), *Choristoneura hebenstreitella* (Müller) (1), *Ptycholoma lecheana* (Linnaeus) (3), *Tortrix viridana* (Linnaeus) (12; includes 7 from one concerted survey of this host) (Tortricidae), and *Pieris brassicae* (Linnaeus) (1) (Pieridae). One specimen mounted with a cocoon of *Hyposoter tricolor* (Ratzeburg), an ichneumonid parasitoid of *Abraxas*, suggests that it may sometimes attack ichneumonoid cocoons although its highly specialised ovipositor seems relatively unsuited to this.

Tribe DELOMERISTINI

This is a rather heterogeneous assemblage and it seems probable that it is an artificial, polyphyletic group. It comprises six genera which are associated by the absence of a basal tooth on the fore tarsal claw of the female; an unusually elongate male subgenital plate; and an internal mandibular tooth and well developed hypostoma in the final instar larva. Adults have widely differing facies, for example, *Theronia* resembles the Pimplini, *Pseudorhyssa* the Rhyssini, and *Delomerista, Hybomischos* and *Perithous* the Ephialtini. The sixth genus, *Atractogaster*, occurs in Europe but not in Britain.

Genus **THERONIA** Holmgren

Theronia is a very large pantropical genus with relatively few species in temperate regions. Two species occur in Europe and one has been found in Britain.

— Large, fore wing length 6·9–12·3 mm. Predominantly reddish orange in colour. Tarsal claws as in Fig. 58. Propodeum without a carina between the area basalis and area superomedia. Gaster virtually impunctate, highly polished. Ovipositor–hind tibia index about 1·0. Hind femur with a ventral longitudinal ridge (not present in any other species of the genus).
· **atalantae** (Poda)

T. atalantae. Very rare or possibly extinct; 7 specimens examined, only two from named localities: England: Kent, Ramsgate, ix.1891 (see Morley, 1909), and Merseyside, Freshfield, 2.ix.1954. Another British specimen was mentioned by Harwood (1921).

We have no direct knowledge of its biology in Britain but it is known in Europe (Aubert, 1969) as a parasitoid of a range of medium-sized to large lepidopterous pupae (either cocooned or naked) situated more or less in the field layer or on shrubs, and also very frequently as a cleptoparasite or hyperparasite, both within Lepidoptera pupae and by attacking ichneumonid cocoons.

Genus DELOMERISTA Foerster

This is a moderate-sized genus, Holarctic in distribution, with about nine species in the western part of the Palaearctic region. Three occur in Britain. Superficially, *Delomerista* look rather like *Scambus*, but the propodeum has a more or less well-defined area superomedia and the gaster is more finely sculptured.

Previously there have been two species of *Delomerista* on the British List. Kloet & Hincks' list (1945) included *D. mandibularis* and *D. pfankuchi*, the latter having been added by Carr (1924). Fitton (*in* Kloet & Hincks, 1978) deleted *pfankuchi* because Carr's collection had been shown to include non-British material (Perkins, 1953) and no other British specimens were then known. *D. laevis* was added to the list on the basis of specimens in the BMNH collection determined by J.F. Perkins and in anticipation of the publication of this handbook. However, recent work on the taxonomy of *Delomerista* by Kasparyan (1977) and Gupta (1982a) shows that this identification is wrong and that the specimens concerned are *D. novita*. Also, British specimens of *D. pfankuchi* have been found and it is reinstated in the British List.

Literature records indicated that at least some *Delomerista* species parasitise the cocoons of sawflies and ichneumonoids, but interpreting these records is rendered even more difficult than usual by the history of confused taxonomy in the genus.

The general colour pattern of the three British species is as follows: mainly black; female with malar space and mandibles, male with face, yellow; legs reddish, hind tibia and tarsi infuscate.

Key to species

1   Female. Tegula mainly blackish. Upper valve of ovipositor rounded at apex (Fig. 160). (Ovipositor–hind tibia index 1·8–2·0. Fore wing length 5·3–8·2 mm.) . . . .
. . . . **mandibularis** (Gravenhorst) (female)

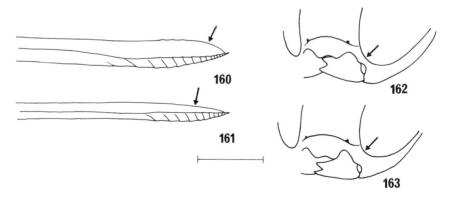

Figs 160–163. *Delomerista* species. 160, apex of ovipositor, lateral view, *D. mandibularis*. 161, apex of ovipositor, lateral view, *D. novita*. 162, malar space, *D. novita* male. 163 malar space, *D. mandibularis* male. Scale line represents 0·5 mm.

— Female or male. Tegula mainly yellowish. Female: upper valve of ovipositor simply tapered at apex (Fig. 161).. 2
2 Gastral tergites distinctly and closely punctate-reticulate, at least in anterior half. (Ovipositor–hind tibia index 1·5. Fore wing length 7·2–7·3 mm.) . .   **pfankuchi** (Brauns)
— Gastral tergites irregularly granulate. 3
3 Female. (Ovipositor–hind tibia index 1·5–1·6. Fore wing length 8·5–9·5 mm.) . . . . .
  . . . . **novita** (Cresson) (female)
— Male. 4
4 Malar space extremely narrow, as in Fig. 162, and always blackish. . . . .
  . . . . **novita** (Cresson) (male)
— Malar space less narrow, as in Fig. 163, and often yellowish.. . . . .
  . . . . **mandibularis** (Gravenhorst) (male)

D. mandibularis. Rare (39 specimens from 7 localities); England: Devon, Oxfordshire, Suffolk, Lincolnshire and Cheshire. Flight period: v–vii. We have seen no reared material.

D. novita. Rare (5 specimens from 4 localities); England: Devon and Cheshire, and Ireland: Down and Wicklow. Flight period: vi, ix. We have seen no reared material.

D. pfankuchi. Rare (2 specimens from 1 locality); England: Oxfordshire. Flight period: vi. The only reared specimen seen emerged from a cocoon of the braconid *Aleiodes grandis* Giraud in the mummified skin of *Amphipyra ?berbera* Rungs (Noctuidae) collected on an oak twig in April.

Genus **PERITHOUS** Holmgren

This is a small genus with species in the Holarctic and Oriental regions. It is not clear from the literature (Aubert, 1969; Gupta 1982b; Oehlke, 1967) how many species there are in Europe; but there are at least three, two of which occur in Britain. Host names in the literature include gall-makers and wood-borers, but all *Perithous* are probably exclusively associated with aculeate Hymenoptera, including those that make subsequent use of such sites. There are a few literature records from Eumenidae and bees,

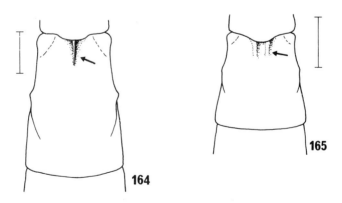

Figs 164–165. *Perithous* species, tergite 2 of gaster, dorsal view. 164, *P. scurra*. 165, *P. divinator*. Scale lines represent 0·5 mm.

but it seems clear that the British species, at least, specialise on sphecid hosts. Both species pupate naked in the host's cell without making a cocoon. The adults are richly marked with red and yellow so that only the gaster looks predominantly black.

Key to species

1 Gaster tergite 2 with a distinct median longitudinal keel anteriorly (Fig. 164), and at least 1·4 (female) or at least 1·5 (male) times as long as broad. Ovipositor–hind tibia index 4·5–5·0. (Fore wing length 6·7–10·5 mm.) . . . . **scurra** (Panzer)
— Gaster tergite 2 without a median keel, occasionally with a pair of vestigial lateromedian keels anteriorly (Fig. 165), at most 1·3 (female) or at most 1·4 (male) times as long as broad. Ovipositor–hind tibia index 2·5–3·2. (Fore wing length 3·7–7·2 mm.) . . . . **divinator** (Rossius)

P. divinator. Moderately common; widely distributed in England as far north as Cumbria, and one specimen seen from Ireland (Cavan), but none from Wales or Scotland. Flight period: vi and vii–viii, bivoltine. Commonly reared from the nests of aculeate Hymenoptera (in particular of Sphecidae) in woody stems or stem-like structures. Oviposition is through the stem, so most or all of the cells in a nest may be parasitised. Danks (1971) recorded this species as a regular prepupal parasitoid of the sphecid wasps *Pemphredon (Cemonus) lethifer* (Shuckard), *P. (C.) inornatus* Say (as *schuckardi* (Morawitz)), and less frequently *Passaloecus singularis* Dahlbohm (as *gracilis* (Curtis), but presumably a misidentification) and *Psenulus schencki* (Tournier), nesting in *Rubus* and other pithy stems, saying also that it is sometimes a secondary parasitoid through *Omalus auratus* (Linnaeus) and *Chrysis (Trichrysis) cyanea* (Linnaeus). In addition to material from the foregoing hosts in *Rubus* and *Rosa* stems, and many specimens reared from such stems without clear host data, we have seen 11 specimens reared from *?Rhopalum (Corynopus) coarctatum* (Scopoli) in *Rubus* stems, 5 from undetermined aculeate hosts in twigs of *Fraxinus*, one from the colletine bee *Hylaeus communis* Nylander, and a further specimen reared from a (presumably superseded) gall of the tortricid moth *Petrova resinella* (Linnaeus) on *Pinus*.

P. scurra. This species is referred to in virtually all recent literature as *mediator*. The synonymy and priority of *scurra* was established by Horstmann (1982).
Moderately common; widely distributed in England and Scotland, and found also in Ireland, but so far not in Wales. Flight period: v–viii and perhaps bivoltine in the south; vi–vii and apparently univoltine in the north. Much less commonly reared than *P. divinator* but probably only because it attacks hosts nesting in less easily gathered substrates. In particular, adults are frequently collected on or around standing timber with soft rotten wood. In addition to a few 'rearings from unknown hosts in such substrates we have seen material reared from Sphecidae determined as *Pemphredon lugubris* (Fabricius) (4 specimens) and *Pemphredon* sp. (1), and one specimen labelled 'ex stump containing *Osmia leucomela*' which we consider a more doubtful indication of host.

Genus **HYBOMISCHOS** Baltazar

Hybomischos used to be treated as a subgenus of *Perithous* but it has recently been accorded generic status (for example, by Gupta, 1982c). The single British species is Holarctic in distribution. The few other species in the genus occur in the Palaearctic and Oriental regions.

— Gaster tergite 1 with anterior three-quarters strongly coriaceous. Hind tarsus whitish with each segment black distally. (Fore wing length 5·7–7·4 mm. Ovipositor–hind tibia index 5·9–7·2. Coloration broadly similar to species of *Perithous*.) . **septemcinctorius** (Thunberg)

H. septemcinctorius. Uncommon, but widely distributed in southern England as far north as Cambridgeshire, and one specimen seen from Ireland (Kildare). Probably commonest in wetlands. Flight period: vii–ix. The only reared specimens seen have been from a rotten oak twig (1 specimen), from straw thatch in which several sphecids were nesting (1), and (on the continent) from *Psenulus fuscipennis* (Dahlbom) (1). The literature records collected by Aubert (1969) suggest that, like *Perithous* species, *H. septemcinctorius* may specialise on sphecid nests, though in the present case the substrates searched are not clear.

Genus **PSEUDORHYSSA** Merrill

Pseudorhyssa comprises just three species: *P. maculicoxis* (Kriechbaumer), which is Holarctic in distribution; *P. alpestris*, which has been found in north-western and central Europe and Japan; and *P. acutidentata* Kusigemati, known only from Japan.

— Fore wing length 5·4–14·1 mm. Mainly black with legs reddish, (hind tibia and tarsus infuscate). Female: face with obscure brownish marks, tergites 1–3 of gaster usually with a redbrown mark near hind margin. Ovipositor–hind tibia index 4·5–5·1. Male: face yellow. **alpestris** (Holmgren)

P. alpestris. Rare (49 specimens from 6 localities); England: Gloucestershire, Hampshire, Berkshire, Oxfordshire, Hertfordshire. Flight period: v–vi. Most of the material seen has been reared from the woodwasp *Xiphydria camelus* (Linnaeus). It is known (Skinner & Thompson, 1960) to be a cleptoparasite, using the oviposition drill hole left by *Rhyssella approximator* (see page 91) to reach the host larva. Unlike Rhyssini, *Pseudorhyssa* species have feeble ovipositors, incapable of drilling through wood. The non-British species *P. maculicoxis* is much better documented and behaves similarly towards siricid hosts already attacked by *Rhyssa persuasoria* (Spradbery, 1969; 1970a).

Tribe **POEMENIINI**

This fairly small tribe has seven genera, three of which are found in Britain. In addition, *Neoxorides*, the fourth European genus, has been recorded as British in error (see p. 17). At least some of the species are most often collected in association with dead standing timber, and all seem likely to attack wood-inhabiting hosts. However, exact host relations are extremely hard to establish in such substrates, not least because the feeding sites of xylophagous insects are so often used subsequently as nesting or pupation sites by entirely different groups.

Morphologically poemeniines resemble species of the subfamily Xoridinae, with which they used to be classified. The two groups can be distinguished easily: in the hind wing of xoridines the abscissa of Cu between M + Cu and cu-a is about as long as cu-a whereas in poemeniines it is much shorter.

On four occasions recently (Carlson, 1979; Gauld, 1984; Gupta, 1985; 1987) this tribe has been referred to by the junior synonym Neoxoridini. Gauld's reference was an inadvertent error. Neither Carlson nor Gupta adduce evidence to counter the arguments of Fitton & Gauld (1976) that under the *International Code* Poemeniini is the valid name.

Genus **POEMENIA** Holmgren

A small Holarctic genus with four species in Europe. Three of these have been taken in Britain, two only very recently. The genus is the only one in the tribe with bidentate mandibles, the lower tooth being rather elongate. The species are black with the legs mainly reddish but (especially in *P. notata* and *P. collaris*) may have extensive cream and red marks.

Key to species

1   Segment 1 of gaster with sternite ending in front of midlength, about level with spiracle (Fig. 166). Fore wing with 3rs-m absent. Thorax extensively red-marked. (Fore wing length 4·0–6·3 mm. Ovipositor–hind tibia index 2.1–2·3.) . . . **notata** Holmgren
—   Segment 1 of gaster with sternite ending behind midlength, well behind level of spiracle (Fig. 167). Fore wing with 3rs-m often present. Thorax sometimes with cream marks, but unless the cream is extensive not with red marks. 2
2   Area of mesopleuron beneath the insertion of the fore wing smooth and shining between the punctures. Pronotum with at least a dorsolateral cream stripe (in front of the tegula). (Fore wing length 3·8–6·9 mm.) . . . . **collaris** (Haupt)
—   Area of mesopleuron beneath the insertion of the fore wing with obsolete rugose-punctate sculpture. Pronotum with at most ventrolateral edge and a spot in front of the tegula cream. (Fore wing length 4·5–10·5 mm. Ovipositor–hind tibia index 1·9–2·4.) . . . .
. . . . **hectica** (Gravenhorst)

P. collaris. Only one male specimen has been examined; the wing length range and female characters are taken from Oehlke (1966).

A single British record; England: Kent, Aylesford, bred from a piece of dead elder containing nests of *Passaloecus eremita* Kohl (Hymenoptera: Sphecidae) and beetle borings (?*Ptilinus pectinicornis* (Linnaeus)), emerged ii.1981. On the continent, Westrich (1980) has reared it in numbers (see Schmidt & Zmudzinski, 1983) from the sphecid *Passaloecus corniger* Shuckard in trap nests.

P. hectica. Rare (13 specimens from 9 localities); England: Devon, Dorset, Hampshire, Gloucestershire; Ireland: Killarney. Flight period: vi–viii. We have seen material collected in association with conifers as well as dead angiosperm trees, but we know of no host records.

P. notata. Only females have been examined: characters of the male are taken from Oehlke (1966).

A single British record; England: Kent, Bedgebury Forest, 3.ix.1980; a female was observed probing with her ovipositor the resin plugs made by *Passaloecus eremita* Kohl

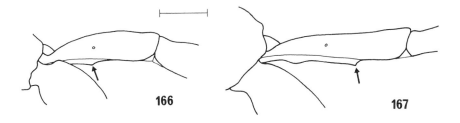

166

167

Figs 166–167. *Poemenia* species, segment 1 of gaster, lateral view. 166, *P. notata* female. 167, *P. hectica* female. Scale line represents 0·5 mm.

(Hymenoptera: Sphecidae) to seal beetle holes containing cells in a tree trunk (G.H.L. Dicker, pers. comm.). On the continent, Westrich (1980) reared a long series (see Schmidt & Zmudzinski, 1983) from trap-nests inhabited by *P. eremita* and, much less often, from those containing *P. corniger* Shuckard.

Genus DEUTEROXORIDES Viereck

This genus comprises only two Palearctic species, one of which occurs in Europe and has been taken in Britain.

— Fore wing length 5·1-11·6 mm. Main sculpture of dorsal half of temple consisting of minute, even, scale-like ridges. Ovipositor–hind tibia index 1·0–1·3. Mainly black with legs yellowish to red-brown, hind tibia infuscate, middle segments of hind tarsus sometimes pale; female antenna sometimes white banded and distally orange-brown; male face yellow. Vein 3rs-m absent from fore wing. . . . . **elevator** (Panzer)

D. elevator. In the past this species has been known by the junior name *albitarsus* (Gravenhorst). The synonymy was established by Horstmann (1982).

Rare; England: Devon, Hampshire, Hereford & Worcester, Surrey, Oxfordshire, Buckinghamshire; Ireland: Wicklow. Flight period: v–vii. The reared material we have seen gives no hint of anything but beetle hosts on the data labels: *Molorchus minor* (Linnaeus) (1 specimen), *Aromia moschata* (Linnaeus) (1), and *Mesites tardii* (Curtis) (1). Other material has been reared from the wood of conifers, from *Alnus*, and collected on dead standing *Quercus* believed to be infested with *Clytus arietis* (Linnaeus) and *Leiopus nebulosus* (Linnaeus).

Genus PODOSCHISTUS Townes

This small Holarctic genus comprises four species, only one of which occurs in Europe.

— Fore wing length 5·4–12·5 mm. Main sculpture of dorsal half of temple consisting of large, uneven, scale-like ridges (Fig. 21). Ovipositor–hind tibia index 2·0–2·3. Mainly black with scutellum and postscutellum yellow marked, legs reddish to yellowish, hind tibia and tarsus infuscate; male face yellow. . . . . **scutellaris** (Desvignes)

P. scutellaris. Rare (9 specimens from 7 localities, and 6 without data); England: East Sussex, Gloucestershire, Oxfordshire, Hertfordshire, Kent, Shropshire; Wales: Gwynedd. Flight period: vi. We have seen no reared material, but it has been collected on standing dead *Quercus* believed to be infested with the beetles *Clytus arietis* (Linnaeus) and *Leiopus nebulosus* (Linnaeus).

Tribe RHYSSINI

This is a moderately large tribe comprising nine genera. It is apparently holophyletic and the main autapomorphic character is the truncate 'horn' on the last gastral tergite of females. Some of the largest ichneumoids are rhyssines and females are particularly noticed because of their very long ovipositors. The Holarctic genus *Megarhyssa*, represented in Europe but not found in Britain, includes the largest and most spectacular species. The majority of rhyssines inhabit montane tropical forest and only three genera are found in Europe. Two species representing two genera occur in Britain. Both are associated with woodwasps.

Genus RHYSSA Gravenhorst

A Holarctic genus with a moderate number of species, two of which occur in Europe. The single British species is well known because it it frequently illustrated as a 'representative' ichneumon-fly in general books on entomology. The female is quite spectacular and can be up to 100 mm from antenna tip to ovipositor apex. However, it is not a common insect and probably receives more attention than it deserves.

— Fore wing length 5·8–25·5 mm. Mid trochantellus simple. Female with the ovipositor guides (tubercles) about halfway along the gastral sternites (2–4) (Fig. 17). Ovipositor–hind tibia index 3·6–4·7. Black species with red legs, and usually conspicuous yellow or cream marks on thorax and propodeum and hind margins of gastral tergites.. **persuasoria** (Linnaeus)

R. persuasoria. Uncommon, although well represented in collections and widely distributed in Great Britain and Ireland. Flight period:v–vii. An extremely well-known parasitoid of a range of economically important siricid woodwasps inhabiting coniferous trees, perhaps detecting infested trees by responding to the woodwasp's fungal symbiont, *Amylostereum* (Spradbery & Kirk, 1978). It is able to develop as a facultative hyperparasite on other parasitoids of siricids (Hanson, 1939), and it seems possible that other wood boring insects present in trees infested by siricids may occasionally be used. The early stages were described by Spradbery (1970a).

Genus RHYSSELLA Rohwer

A small Holarctic genus with two European species. The dividing line between *Rhyssella* and *Megarhyssa* is rather vague and the main justification for their continued recognition as separate genera would seem to be that they attack different groups of hosts: Xiphydriidae in the case of *Rhysella* and Siricidae in the case of *Megarhyssa*, although even this is not established beyond doubt. Only one species occurs in Britain.

— Fore wing length 5·4–15·6 mm. Mid trochantellus with a small lamella antero-ventrally at its apex (Fig. 16). Female with the ovipositor guides (tubercles) near the anterior margins of the gastral sternites (2–4) (Fig. 18). Ovipositor–hind tibia index 3·8–4·3. Male with the posterior margins of gastral tergites 3–6 excised (Fig. 19). Mainly black species with legs red. . . . . **approximator** (Fabricius)

R. approximator. Uncommon; widely distributed in England and Scotland but there are no records from Wales or Ireland. Flight period: v–viii. The woodwasp genus *Xiphydria* (or Xiphydriidae) has been the only host name seen on the labels of reared specimens (47). *R. approximator* has been reared about equally from our two commonest *Xiphydria* species: the widespread *X. camelus* (Linnaeus) infesting the wood of *Alnus* and *Betula*, and the more restricted *X. prolongata* (Geoffroy) which occurs, sometimes at great density, in *Salix*. It would be surprising if it did not also attack *X. longicollis* (Geoffroy), as is recorded on the continent. Chrystal & Skinner (1932) gave an account of its biology (as *Thalessa curvipes* (Gravenhorst)).

Tribe DIACRITINI

This small tribe comprises only four described species in two genera, one of which occurs in Europe. Structurally they are rather aberrant and nothing is known of their host associations or life-histories, apart from a series of four specimens (in the BMNH and possibly representing a further, undescribed genus) from Java, labelled as being reared from the leaf mining chrysomelid beetle *Promecotheca nucifera* Maulik (a pest of coconut). Their inclusion in the Pimplinae should be regarded as provisional: some workers (e.g. Perkins, 1940) place *Diacritus* in the Oxytorinae.

Genus **DIACRITUS** Foerster

Diacritus is Holarctic in distribution with one species in Europe.

— Fore wing length 4·5–5·6 mm. Ovipositor–hind tibia index 0·6–0·7. Notauli long and deep and meeting posteriorly. Sternite of segment 1 of gaster as in Fig. 22. Pronotum with a deep mediodorsal longitudinal groove. Fore wing with areolet petiolate (that is veins 2rs-m and 3rs-m meeting and fusing before they meet Rs). Black with hind margins of tergites (except first) of gaster and proximal parts of fore and mid legs whitish; antenna, mouthparts and remainer of legs brownish; male with more extensive white marks: on mouthparts, face and pronotum. . . . . **aciculatus** (Vollenhoven)

D. aciculatus. Rare, but widely distributed in southern England north to Norfolk; and also found in Scotland: Angus, and in Ireland: Wicklow. Flight period: vii–ix. There are no host records.

Acknowledgements

We are indebted to the following for access to, and information concerning, the collections under their care: B. R. Baker, L. R. Cole, J. P. O'Connor, W. A. Ely, W. A. Foster, T. Kronestedt, G. C. McGavin, C. Johnson, R. Nash, M. Reilly, and the late G. C. Varley. By donating (mainly to NMS) recently collected and especially reared material the following have contributed greatly to the biological and distributional details we have been able to give: N. P. Ashmole, R. R. Askew, M. C. Askins, J. S. Badmin, S. I. Baldwin, B. D. Batty, K. P. Bland, E. S. Bradford, M. R. Britton, J. P. Brock, P. W. Brown, K. Catley, P. J. Chandler, I. C. Christie, M. J. W. Cock, S. G. Compton, A. E. Cooper, M. F. V. Corley, S. Crellin, P. W. Cribb, J. E. Dalingwater, G. H. L. Dicker, C. M. Drake, G. R. Else, R. E. Evans, C. Felton, R. L. E. Ford, A. P. Foster, G. N. Foster, C. Geddes, H. J. C. Godfray, J. L. Gregory, D. J. R. Haigh, N. M. Hall, P. R. Harvey, B. A. Hawkins, R. J. Heckford, D. Horsfield, I. R. Hudson, R. J. T. Jarvis, L. M. Jones-Walters, S. Judd, M. A. Kirby, R. P. Knill-Jones, S. Koptur, J. R. Langmaid, A. D. Liston, S. C. Littlewood, R. I. Lorimer, E. J. Lovesey, B. M. Lyszkowski, I. MacGowan, H. Mendel, J. E. D. Milner, F. M. Murphy, J. M. Nelson, J. A. Owen, B. T. Parsons, J. H. Payne, E. C. Pelham-Clinton, M. G. M. Randall, G. E. Rotheray, M. C. & W. Shaw, D. A. Sheppard, P. R. Shirley, A. N. B. Simpson, R. A. Softly, P. A. Sokoloff, P. H. Sterling, J. A. Stewart, N. E. Stork, W. A. Thornhill, C. R. Vardy, W. A. Watson, A. J. White, S. A. & D. C. Wilkinson, D. W. Yalden, M. R. Young and B. Zonfrillo. We are grateful to Isobel Baldwin, David Carter, Peter Merrett, Mike Morris and John Owen for their help with identifying hosts. Francesca Shaw has jumped through many hoops in support of fieldwork with exceptionally good grace, and the copious supply of *Drosophila* cultures provided by Sandra Grant, Department of Zoology, University of Edinburgh, has enabled well over 150 polysphinctines to be reared. Paul Eggleton and Neil Springate made a useful contribution by testing keys and commenting on them and other sections of the handbook. The Royal Entomological Society and CAB International Institute of Entomology kindly gave permission for us to reproduce some of the figures from Eady (1968).

References

ADOLFSSON, J. 1984. Tallspinnaren och dess parasitoider — ett känsligt samspel. *Entomologisk Tidskrift* **105**: 15–24.

ALLIONI, C. 1766. Manipulus insectorum Taurinensium. *Mélanges de Philosophie et de Mathematique de la Société Royale de Turin* **1762–1765** (3): 185–198.

ARTHUR, A P. 1966. Associative learning in *Itoplectis conquisitor* (Say) (Hymenoptera: Ichneumonidae). *Canadian Entomologist* **98**: 213–233.

ARTHUR A. P. 1981. Host acceptance by parasitoids. *In* Nordlund, D. A., Jones, R. L. & Lewis, W. J. (Eds) *Semiochemicals, their role in pest control*: 97–120. New York.

ARTHUR, A. P. & WYLIE, H. G. 1959. Effects of host size on sex ratio, development time and size of *Pimpla turionellae* (L.) (Hymenoptera: Ichneumonidae). *Entomophaga* **4**: 297–301.

ASKEW R. R. & SHAW, M. R. 1986. Parasitoid communities: their size, structure and development. *In* Waage, J. & Greathead, D. (Eds) *Insect parasitoids*: 225–264. London.

AUBERT, J. F. 1959. Biologie de quelques Ichneumonidae Pimplinae et examen critique de la théorie de Dzierzon. *Entomophaga* **4**: 75–188.

AUBERT, J. F. 1966. Les Ichneumonides *Scambus* Htg., *Acropimpla* Townes et *Iseropus* Först. du Musée zoologique de Lausanne, avec clefs inédites pour toutes les espèces européennes. *Mitteilungen der Schweizerischen Entomologischen Gesellschaft* **38**: 145–172.

AUBERT, J. F. 1967. Supplément à la révision des Ichneumonides *Scambus* Htg. ouest-paléarctiques. *Mitteilungen der Schweizerischen Entomologischen Gesellschaft* **40**: 56–62.

AUBERT, J. F. 1969. *Les Ichneumonides ouest-paléarctiques et leurs hôtes. 1. Pimplinae, Xoridinae, Acaenitinae*. 302pp. Paris.

AUBERT, J. F. 1970. Noveau supplément aux Ichneumonides non pétiolées, avec description d'un genre nouveau. *Bulletin de la Société Entomologique de Mulhouse* **1970**: 49–56.

BIGNELL, G. C. 1898. The Ichneumonidae (parasitic flies) of the south of Devon. *Transactions of the Devonshire Association for the Advancement of Science, Literature and Art* **30**: 458–504.

BRIDGMAN, J. B. 1886. Further additions to the Rev. T. A. Marshall's catalogue of British Ichneumonidae. *Transactions of the Entomological Society of London* **1886**: 335–373.

BRUZZESE, E. 1982. Observations on the biology of *Pseudopimpla pygidiator* Seyrig (Hym., Ichneumonidae), a parasite of the blackberry stem boring sawfly *Hartigia albomaculatus* (Stein) (Hym., Cephidae). *Entomologist's Monthly Magazine* **118**: 249–252.

CAMPBELL, R. W. 1963. Some ichneumonid-sarcophagid interactions in the gypsy moth *Porthetria dispar* (L.) (Lepidoptera: Lymantriidae). *Canadian Entomologist* **95**: 337–345.

CARLETON, M. 1939. The biology of *Pontania proxima* Lep., the bean gall sawfly of willows. *Journal of the Linnean Society* (Zoology) **40**: 575–624.

CARLSON, R. W. 1979. Ichneumonidae. *In* Krombein, K. V., Hurd, P. D., Smith, D. R. & Burks, B. D. (Eds). *Catalog of Hymenoptera in America North of Mexico*. **1**: 315–740. Washington.

CARR, L. A. 1924. The Ichneumonidae of the Lichfield district, Staffordshire. *Transactions of the North Staffordshire Field Club* **58** (Appendix): 1–70.

CARTON, Y. 1978. Biologie de *Pimpla instigator* (Hym.: Ichneumonidae) IV. Modalités du développement larvaire en fonction du site de ponte; rôle des réactions hémocytaires de l'hôte. *Entomophaga* **23**: 249–259.

CHRIST, J. L. 1791. *Naturgeschichte, Klassification und Nomenklatur der Insekten vom Bienen, Wespen und Ameisengeschlecht*. 535 pp. and 60 plates. Frankfurt am Main.

CHRYSTAL, R. N. & SKINNER, E. R. 1932. Studies on the biology of the woodwasp *Xiphydria prolongata* Geoffr. and its parasite *Thalessa curvipes* Grav. *Scottish Forestry Journal* **46**: 36–51.

CHVÁLA, M., DOSKOČIL, J., MOOK, J. H. & POKORNÝ, V. 1974. The genus *Lipara* Meigen (Diptera, Chloropidae), systematics, morphology, behaviour, and ecology. *Tijdschrift voor Entomologie* **117**: 1–25.

CLARET, J. 1973. La diapause facultative de *Pimpla instigator* (Hym.: Ichneumonidae) I. Rôle de la photopériode. *Entomophaga* **18**: 409–418.

CLARET, J. 1978. La diapause facultative de *Pimpla instigator* (Hym.: Ichneumonidae) II. Rôle de la température. *Entomophaga* **23**: 411–415.

CLARET, J. & CARTON, Y. 1975. Influence de l'espèce-hôte sur la diapause de *Pimpla instigator* F. (Hyménoptère, Ichneumonidae). *Compte Rendu de l'Academie des Sciences. Paris* (D) **281**: 279–282.

COLE, L. R. 1959. On the defences of lepidopterous pupae in relation to the oviposition behaviour of certain Ichneumonidae. *Journal of the Lepidopterists' Society* **13**: 1–10.

COLE, L. R. 1967. A study of the life-cycles and hosts of some Ichneumonidae attacking pupae of the green oak-leaf roller moth, *Tortrix viridana* (L.) (Lepidoptera: Tortricidae) in England. *Transactions of the Royal Entomological Society of London* **119**: 267–281.

COLE, L. R. 1981. A visible sign of fertilisation action during oviposition by an ichneumonid wasp, *Itoplectis maculator*. *Animal Behaviour* **29**: 299–300.

CONSTANTINEANU M. I. & PISICA, C. 1977. Hymenoptera, Familia Ichneumonidae, Subfamiliile Ephialtinae, Lycorininae, Xoridinae si Acaenitinae. *Fauna Republicii Socialiste România* (Insecta) **9** (7): 1–305.

CUSHMAN, R. A. 1926. Address of the retiring President. *Proceedings of the Entomological Society of Washington* **28**: 25–51.

CUSHMAN. R. A. 1938. A new European species of *Epiurus*, parasitic on a leafmining sawfly (Hymenoptera: Ichneumonidae). *Journal of the Washington Academy of Sciences* **28**: 27–28.

DALLA TORRE, C. G. de 1901. *Catalogus Hymenopterorum* **3**: 1–544. Lipsiae.

DANKS, H. V. 1971. Biology of some stem-nesting aculeate Hymenoptera. *Transactions of the Royal Entomological Society of London* **122**: 323–399.

DEGEER, C. 1771. *Memoires pour servir a l'histoire des insects*. **2** (2): 617–1175. Stockholm.

DOUTT, R. L. 1964. Biological characteristics of entomophagous adults. *In* DeBach, P. *Biological Control of Insect Pests and Weeds:* 145–167. London.

DOWDEN, P. B. 1941. Parasites of the birch leaf-mining sawfly (*Phyllotoma nemorata*). *Technical Bulletin, United States Department of Agriculture* **757**: 1–55.

EADY. R. D. 1968. Some illustrations of microsculpture in the Hymenoptera. *Proceedings of the Royal Entomological Society of London* (A) **43**: 66–72.

EVANS, H. F. 1985. Great spruce bark beetle, *Dendroctonus micans*: an exotic pest new to Britain. *Antenna* **9**: 117–121.

FABRICIUS, J. C. 1793. *Entomologica Systematica*. **2**, 519 pp. Hafniae.

FINLAYSON, T. 1967. A classification of the subfamily Pimplinae (Hymenoptera: Ichneumonidae) based on final instar larval characteristics. *Canadian Entomologist* **99**: 1–8.

FITTON, M. G. 1976. The Western Palaeartic Ichneumonidae (Hymenoptera) of British authors. *Bulletin of the British Museum (Natural History)* (Entomology) **32**: 301–373.

FITTON, M. G. 1981. The British Acaenitinae (Hymenoptera: Ichneumonidae). *Entomologist's Gazette* **32**: 185–192.

FITTON, M. G. & GAULD, I. D. 1976. The family-group names of the Ichneumonidae (excluding Ichneumoninae) (Hymenoptera). *Systematic Entomology* **1**: 247–258.

FITTON, M. G., SHAW, M. R. & AUSTIN, A. D. 1987. The Hymenoptera associated with spiders in Europe. *Zoological Journal of the Linnean Society* **90**: 65–93.

FOURCROY, A. F. de (Ed.) 1785. *Entomologia Parisiensis*. **2**, pp. 233–544. Parisiis.

FÜHRER, E. 1975. Über die physiologische Spezifität des polyphagen Puppenparasiten *Pimpla turionellae* L. (Hym., Ichneumonidae) und ihre ökologischen Folgen. *Zentralblatt für das gesamte Forstwesen* **92**: 218–227.

FÜHRER, E. & KILINCER, N. 1972. Die motorische Aktivität der endoparasitischen Larven von *Pimpla turionellae* L. und *Pimpla flavicoxis* Ths. in der Wirtspuppe. *Entomophaga* **17**: 149–165.

FÜHRER, E. & WILLERS, D. 1986. The anal secretion of the endoparasitic larva *Pimpla turionellae:* sites of production and effects. *Journal of Insect Physiology* **32**: 361–367.

GAULD, I. D. 1984. The Pimplinae, Xoridinae, Acaenitinae and Lycorininae (Hymenoptera: Ichneumonidae) of Australia. *Bulletin of the British Museum (Natural History)* (Entomology) **49**: 235–339.

GAULD. I. D. & FITTON, M. G. 1981. Keys to the British xoridine parasitoids of wood-boring beetles (Hymenoptera: Ichneumonidae). *Entomologist's Gazette* **32**: 259–267.

GAULD, I. D. & MOUND, L. A. 1982. Homoplasy and the delineation of holophyletic genera in some insect groups. *Systematic Entomology* **7**: 73–86.

GMELIN, J. F. 1790. *In* Linnaeus, C., *Systema Naturae* (13th edn) **1** (5): 2225–3020. Lipsiae.

GRAVENHORST, J. L. C. 1829. *Ichneumonologia Europaea* **3**, 1097pp. Vratislaviae.

GUPTA, V. K. 1982a. A revision of the genus *Delomerista* (Hymenoptera: Ichneumonidae). *Contributions of the American Entomological Institute* **19** (1): 1–42.

GUPTA, V. K. 1982b. A review of the genus *Perithous*, with descriptions of new taxa (Hymenoptera: Ichneumonidae). *Contributions of the American Entomological Institute* **19** (4): 1–20.

GUPTA, V. K. 1982c. A study of the genus *Hybomischos* (Hymenoptera: Ichneumonidae). *Contributions of the American Entomological Institute* **19** (5): 1–5.

GUPTA, V. K. 1985. A review of the tribe Neoxoridini of the World (Hymenoptera: Ichneumonidae: Pimplinae). *Oriental Insects* **19**: 323–329.

GUPTA, V. K. 1987. The Ichneumonidae of the Indo-Australian area (Hymenoptera). *Memoirs of the American Entomological Institute* **41**: 1–1210.

GUPTA, V. K. & TIKAR, D. T. 1969. Taxonomic identity of pimpline genera *Flavopimpla* Betrem and *Afrephialtes* Benoit (Hymenoptera: Ichneumonidae). *Oriental Insects* **3**: 269–278.

GUPTA, V. K. & TIKAR, D. T. 1978. Ichneumonologia Orientalis. Part 1. The tribe Pimplini. *Oriental Insects Monographs* **1** (1976), 312pp. Delhi.

HALSTEAD, A. J. 1987. Unusual behaviour by *Pimpla instigator* (F.) (Hym., Ichneumonidae) *Entomologist's Monthly Magazine* **123**: 189.

HANCOCK, G. L. R. 1925. Notes on the hibernation of Ichneumonidae and on some parasites of *Tortrix viridana* L. *Entomologist's Monthly Magazine* **61**: 23–28.

HANSON, H. S. 1939. Ecological notes on the *Sirex* wood wasps and their parasites. *Bulletin of Entomological Research* **30**: 27–65.

HARWOOD, B. S. 1921. *Theronia atalantae*, Poda, in Britain. *Entomologist* **54**: 148.

HEATH, J. & EMMET, A. M. (Eds) 1985. *The moths and butterflies of Great Britain and Ireland* **2**, 460pp. Colchester.

HORSTMANN, K. 1982. Revision der von Panzer beschriebenen Ichneumoniden-Arten. *Spixiana* **5**: 231–246.

HOUSE, H. L. 1978. An artificial host: encapsulated synthetic medium for *in vitro* oviposition and rearing the endoparasitoid *Itoplectis conquisitor* (Hymenoptera: Ichneumonidae). *Canadian Entomologist* **110**: 331–333.

HUDSON, I. R. 1985. Notes on species of Ichneumonidae reared as ectoparasites of spiders. *Proceedings and Transactions of the British Entomological and Natural History Society* **18**: 32–34.

JACKSON, D. J. 1937. Host-selection in *Pimpla examinator* F. (Hymenoptera). *Proceedings of the Royal Entomological Society of London* (A) **12**: 81–91.

JANZON, L. Å. 1982. Description of the egg and larva of *Euphranta connexa* (Fabricius) (Diptera: Tephritidae) and of the egg of its parasitoid *Scambus brevicornis* (Gravenhorst) (Hymenoptera: Ichneumonidae). *Entomologica Scandinavica* **13**: 313–316.

JUILLET, J. A. 1959. Morphology of immature stages, life-history, and behaviour of three hymenopterous parasites of the European Pine Shoot Moth, *Rhyacionia buoliana* (Schiff.) (Lepidoptera: Olethreutidae). *Canadian Entomologist* **91**: 709–719.

JUSSILA, R. & KÄPYLÄ, M. 1975. Observations on *Townesia tenuiventris* (Hlmgr) (Hym., Ichneumonidae) and its hosts *Chelostoma maxillosum* (L.) (Hym., Megachilidae) and *Trypoxylon figulus* (L.) (Hym., Sphecidae). *Annales Entomologici Fennici* **41**: 81–86.

JUSSILA, R. & NUORTEVA, P. 1968. The ichneumonid fauna in relation to an outbreak of *Oporinia autumnata* (Bkh.) (Lep., Geometridae) on subarctic birches. *Annales Zoologici Fennici* **5**: 273–275.

KASPARYAN, D. R. 1973. Review of Palaearctic ichneumonids of the tribe Pimplini (Hymenoptera, Ichneumonidae). The genera *Itoplectis* Först. and *Apechtis* Först. *Entomologicheskoe Obozrenie* **52**: 665–681. (In Russian) [English translation in *Entomological Review, Washington* **52** (3): 444–455]

KASPARYAN, D. R. 1974. A review of Palaearctic species of the tribe Pimplini (Hymenoptera, Ichneumonidae). The genus *Pimpla* Fabricius. *Entomologicheskoe Obozrenie* **53**: 382–403. (In Russian) [English translation in *Entomological Review, Washington* **53** (2): 102–117]

KASPARYAN, D. R. 1977. Review of the European species of ichneumonids of the genus *Delomerista* Foerster (Hymenoptera, Ichneumonidae). *[New and little known species of insects of the European part of the USSR]*: 69–75. Leningrad. (In Russian)

KASPARYAN, D. R. (Ed.) 1981. [Hymenoptera, Ichneumonidae. *Keys to the insects of the European part of the U.S.S.R.*] **3**: 1–688. (In Russian)

KISHI, Y. 1970. Difference in the sex ratio of the pine bark weevil parasite, *Dolichomitus* sp. (Hymenoptera: Ichneumonidae), emerging from different host species. *Applied Entomology and Zoology* **5**: 126–132.

KLOET, G. S. & HINCKS, W. D. 1945. *A check list of British insects.* 483pp. Stockport.

KLOET, G. S. & HINCKS, W. D. 1972. *A check list of British insects.* (2nd edn) Part 2: Lepidoptera. *Handbooks for the identification of British insects* **11** (2): 1–153.

KLOET, G. S. & HINCKS, W. D. 1976. *A check list of British insects.* (2nd edn) Part 5: Diptera. *Handbooks for the identification of British insects* **11** (5): 1–139.

KLOET, G. S. & HINCKS, W. D. 1977. *A check list of British insects.* (2nd edn) Part 3: Coleoptera. *Handbooks for the identification of British insects* **11** (3): 1–105.

KLOET, G. S. & HINCKS, W. D. 1978. *A check list of British insects.* (2nd edn) Part 4: Hymenoptera. *Handbooks for the identification of British insects* **11** (4): 1–159.

LEIGHTON, A. C. 1972. *Transport and communication in early Medieval Europe, AD500–1100.* 257pp. Newton Abbot.

LEIUS, K. 1960. Attractiveness of different foods and flowers to the adults of some hymenopterous parasites. *Canadian Entomologist* **92**: 369–376.

LEIUS, K. 1961a. Influence of food on fecundity and longevity of adults of *Itoplectis conquisitor* (Say) (Hymenoptera: Ichneumonidae). *Canadian Entomologist* **93**: 771–780.

LEIUS, K. 1961b. Influence of various foods on fecundity and longevity of adults of *Scambus buolianae* (Htg.) (Hymenoptera: Ichneumonidae). *Canadian Entomologist* **93**: 1079–1084.

MASON, W. R. M. 1974. Shipping alcohol collections in plastic bags. *Proceedings of the Entomological Society of Washington* **76**: 229–230.

MERRETT, P., LOCKET, G. H. & MILLIDGE, A. F. 1985. A check list of British spiders. *Bulletin of the British Arachnological Society* **6**: 381–403.

MORLEY, C. 1908. *Ichneumonologia Britannica.* 3, 328pp. London.

MORLEY, C. 1909. *Theronia atalantae,* Poda, as British. *Entomologist* **42**: 65.

MÜLLER, O. F. 1776. *Zoologicae Danicae Prodromus.* 274pp. Havniae.

NIELSEN, E. 1923. Contributions to the life history of the pimpline spider parasites (*Polysphincta, Zaglyptus, Tromatobia*) (Hym. Ichneum.). *Entomologiske Meddelelser* **14**: 137–205.

NIELSEN, E. 1928. A supplementary note upon the life histories of the Polysphinctas (Hym. Ichneum.). *Entomologiske Meddelelser* **16**: 152–155.

NIELSEN, E. 1929. A second supplementary note upon the life histories of the Polysphinctas (Hym. Ichneum.). *Entomologiske Meddelelser* **16**: 366–368.

NIELSEN, E. 1935. A third supplementary note upon the life histories of the Polysphinctas (Hym. Ichneum.). *Entomologiske Meddelelser* **19**: 191–215.

NIELSEN, E. 1937. A fourth supplementary note upon the life histories of the Polysphinctas (Hym. Ichneum.). *Entomologiske Meddelelser* **20**: 25–28.

NOYES, J. S. 1982. Collecting and preserving chalcid wasps (Hymenoptera: Chalcidoidea). *Journal of Natural History* **16**: 315–334.

NUORTEVA, P. & JUSSILA, R. 1967. Seasonal and zonal distribution of Ichneumonidae (Hym.) on a subarctic fell during a calamity of the geometrid moth *Oporinia autumnata* (Bkh.) on birches. *Annales Entomologici Fennici* **33**: 155–163.

NUORTEVA, P. & JUSSILA, R. 1969. Incidence of ichneumonids on a subarctic fell after a calamity of the geometrid moth *Oporinia autumnata* (Bkh.) on birches. *Annales Entomologici Fennici* **35**: 153–160.

OEHLKE, J. 1966. Die westpalaearktischen Arten der Tribus Poemeniini (Hymenoptera: Ichneumonidae). *Beiträge zur Entomologie* **15**: 881–892.

OEHLKE, J. 1967. Westpaläarktische Ichneumonidae I: Ephialtinae. *Hymenopterorum Catalogus* (nova editio) **2**: 1–49.

OPINION 159. 1945. On the status of the names *Ephialtes* Schrank, 1802, *Ichneumon* Linnaeus, 1758, *Pimpla* Fabricius, [1804–1805], and *Ephialtes* Gravenhorst, 1829 (Class Insecta, Order Hymenoptera). *Opinions of the International Commission on Zoological Nomenclature* **2**: 275–290.

OSMAN, S. E. 1978. Der Einfluss der Imaginalernährung und der Begattung auf die Sekretproduktion der wieblichen Genitalanhangdrüsen und auf die Eireifung von *Pimpla turionellae* L. (Hym., Ichneumonidae). *Zeitschrift für angewandte Entomologie* **85**: 113–122.

OWEN, J., TOWNES, H. & TOWNES, M. 1981. Species diversity of Ichneumonidae and Serphidae (Hymenoptera) in an English surburban garden. *Biological Journal of the Linnean Society* **16**: 315–336.

PERKINS, J. F. 1940. Notes on the synonymy of some genera of European Pimplinae (s.l.) (Hym. Ichneumonidae). *Entomologist* **73**: 54–56.

PERKINS, J. F. 1941. A synopsis of the British Pimplini, with notes on the synonymy of the European species (Hymenoptera Ichneumonidae). *Transactions of the Royal Entomological Society of London* **91**: 637–659.

PERKINS, J. F. 1943. Preliminary notes on the synonymy of the European species of the *Ephialtes* complex (Hym.), Ichneumonidae. *Annals and Magazine of Natural History* (11) **10:** 249–273.

PERKINS, J. F. 1946. *Ephialtes diversicostae* Perkins and *Pimpla arctica* Zett. (Hym., Ichneumonidae) new to Britain. *Entomologist's Monthly Magazine* **82:** 206.

PERKINS, J. F. 1952. European, and reputed British, Ichneumonidae in the Sir Joseph Banks collection. *Entomologist* **85:** 66–68.

PERKINS, J. F. 1953. Notes on British Ichneumonidae with descriptions of new species (Hym., Ichneumonidae). *Bulletin of the British Museum (Natural History)* (Entomology) **3:** 103–176.

PERKINS, J. F. 1957. Notes on some Eurasian "*Itoplectis*", with descriptions of new species. *Mitteilungen der Schweizerischen Entomologischen Gesellschaft* **30:** 323–326.

PERKINS, J. F. 1959. Ichneumonidae, key to subfamilies and Ichneumonidae-I. *Handbooks for the identification of British insects* **7** (2ai): 1–116.

PODA, N. 1761. *Insecta Musei Graecensis.* 127pp. Graecii.

RAGHI-ATRI, F. 1980. Über das Vorkommen von *Scambus detrita* Holmgr. als Parasit von *Lipara similis* Schin. in Berlin. *Deutsche Entomologische Zeitschrift* (N.F.) **27:** 185–187.

RETZIUS, A. I. 1783. *Genera et species insectorum.* 220 pp. Lipsiae.

RICHARDS, O. W. 1977. Hymenoptera, introduction and key to families. *Handbooks for the identification of British insects* **6** (1): 1–100.

ROJAS-ROUSSE, D. & BENOIT, M. 1977. Morphology and biometry of larval instars of *Pimpla instigator* (F.) (Hymenoptera: Ichneumonidae). *Bulletin of Entomological Research* **67:** 129–141.

ROSSIUS, P. 1790. *Fauna Etrusca.* **2,** 348pp. Liburni.

SALT, G. 1931. Parasites of the wheat-stem sawfly, *Cephus pygmaeus,* Linnaeus, in England. *Bulletin of Entomological Research* **22:** 479–545.

SANDLAN, K. 1979a. Host-feeding and its effects on the physiology and behaviour of the ichneumonid parasitoid, *Coccygomimus turionellae. Physiological Entomology* **4:** 383–392.

SANDLAN, K. 1979b. Sex ratio regulation in *Coccygomimus turionella* Linneaus (Hymenoptera: Ichneumonidae) and its ecological implications. *Ecological Entomology* **4:** 365–378.

SANDLAN, K. 1980. Host location by *Coccygomimus turionellae* (Hymenoptera: Ichneumonidae). *Entomologia Experimentalis et Applicata* **27:** 233–245.

SCHMIDT, K. & ZMUDZINSKI, F. 1983. Beiträge zur Kenntnis der badischen Schlupfwespen-fauna (Hymenoptera, Ichneumonidae) 1. Xoridinae, Acaenitinae, Pimplinae (Poemeniini, Rhyssini). *Andrias* **3:** 97–103.

ŠEDIVÝ, J. 1963. Die europäischen Arten der Gattungen *Laufeia* Tosq. *Polysphincta* Grav. und *Zatypota* Först. (Hym., Ichneumonidae). *Acta Entomologica Musei Nationalis Pragae* **35:** 243–261.

SHAW, M. R. 1986. *Coleocentrus excitator* (Poda) (Hymenoptera: Ichneumonidae) new to Britain. *Entomologist's Gazette* **37:** 221–224.

SHAW, M. R. & ASKEW, R. R. 1976. Parasites. *In* Heath,J. (Ed.) *The moths and butterflies of Great Britain and Ireland* **1:** 24–56. Oxford.

SHERBORN, C. D. 1902. *Index Animalium* (sectio prima a kalendis Ianuariis, MDCCLVIII usque ad finem Decembris, MDCCC). 1195pp. Cantabrigiae.

SKINNER, E. R. & THOMPSON, G. H. 1960. FILM: *The alder woodwasp and its insect enemies.*

SMITHERS, C. N. 1956. On *Philopsyche abdominalis* Morley (Hym.: Ichneumonidae),a parasite of *Acanthopsyche junodi* Heylaerts (Lep.: Psychidae). *Journal of the Entomological Society of Southern Africa* **19:** 225–249.

SPRADBERY, J. P. 1968. A technique for artificially culturing ichneumonid parasites of wood-wasps (Hymenoptera: Siricidae). *Entomologia Experimentalis et Applicata* **11:** 257–260.

SPRADBERY, J. P. 1969. The biology of *Pseudorhyssa sternata* Merrill (Hym., Ichneumonidae), a cleptoparasite of siricid woodwasps. *Bulletin of Entomological Research* **59:** 291–297.

SPRADBERY, J. P. 1970a. The immature stages of European ichneumonid parasites of siricine woodwasps. *Proceedings of the Royal Entomological Society of London* (A) **45:** 14–28.

SPRADBERY, J. P. 1970b. Host finding by *Rhyssa persuasoria* (L.), an ichneumonid parasite of siricid woodwasps. *Animal Behaviour* **18:** 103–114.

SPRADBERY, J. P. & KIRK, A. A. 1978. Aspects of the ecology of siricid woodwasps (Hymenoptera: Siricidae) in Europe, North Africa and Turkey with special reference to the biological control of *Sirex noctilio* F. in Australia. *Bulletin of Entomological Research* **68:** 341–359.

STELFOX, A. W. 1929. On the distinction of *Pimpla clavicornis* Thoms. and *P. curticauda* Kriech. *Entomologist's Monthly Magazine* **65:** 17–18.

THORPE, W. H. & CAUDLE, H. B. 1938. A study of the olfactory responses of insect parasites to the food plant of their host. *Parasitology* **30:** 523–528.

TOWNES, H. 1969. The genera of Ichneumonidae, part 1. *Memoirs of the American Entomological Institute* **11:** 1–300.

TOWNES, H. 1972. A light-weight Malaise trap. *Entomological News* **83:** 239–247.

TOWNES, H. & CHIU, S. 1970. The Indo-Australian species of *Xanthopimpla* (Ichneumonidae). *Memoirs of the American Entomological Institute* **14:** 1–372.

TOWNES, H., MOMOI, S. & TOWNES, M. 1965. A catalogue and reclassification of the Eastern Palearctic Ichneumonidae. *Memoirs of the American Entomological Institute* **5:** 1–661.

TOWNES, H. & TOWNES, M. 1960. Ichneumon-flies of America North of Mexico: 2 Subfamilies Ephialtinae, Xoridinae, Acaenitinae. *Bulletin of the United States National Museum* **216** (2): 1–676.

VINSON, S. B. 1976. Host selection by insect parasitoids. *Annual Review of Entomology* **21:** 109–133.

VINSON, S. B. 1981. Habitat location. *In* Nordlund, D. A., Jones, R. L. & Lewis, W. J. (Eds) *Semiochemicals, their role in pest control:* 51–77. New York.

WARDLE, A. R. & BORDEN, J. H. 1985. Age-dependent associative learning by *Exeristes roborator* (F.) (Hymenoptera: Ichneumonidae). *Canadian Entomologist* **117:** 605–616.

WESELOH, R. M. 1981. Host location by parasitoids. *In* Nordlund, D. A., Jones, R. L. & Lewis, W. J. (Eds) *Semiochemicals, their role in pest control:* 79–95. New York.

WESTRICH, P. 1980. Die Stechimmen (Hymenoptera Aculeata) des Tübinger Gebiets mit besonderer Berücksichtigung des Spitzbergs. *Veröffentlichungen Naturschutz Landschaftsplege Bad.-Württ.* **51/52:** 601–680.

ZWAKHALS, C. J. 1987. Revision of the genus *Alophosternum* Cushman with a new species from Japan (Hymenoptera: Ichneumonidae, Pimplinae) *Entomologische Berichten* **47:** 108–111.

Appendix

Illustrations of microsculpture

Below are some illustrations of cuticular microsculpture to show the meaning of the terminology used in this handbook. We are including these figures for ease of reference. They are taken directly from Eady's paper (1968), which also has a discussion and further figures of more complex sculpture. The figure numbers are not those used in the original publication.

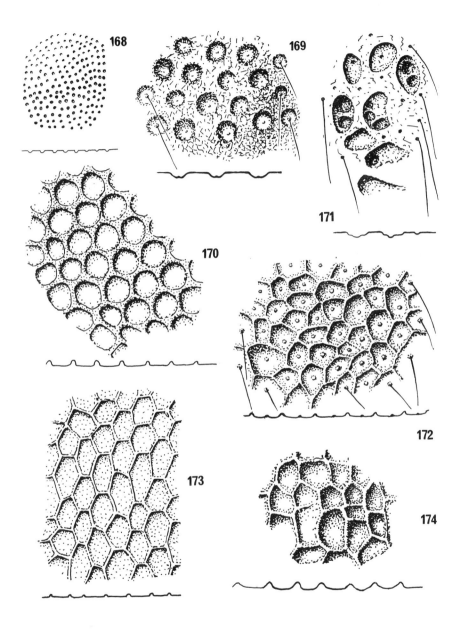

Figs 168–174. Cuticular microsculpture. 168, punctulate. 169, punctate, with hairs. 170, punctate-reticulate. 171, foveolate. 172, rugose-punctate. 173, reticulate. 174, reticulate-rugose. The line below each figure represents an imaginary section of the cuticle. All figures are taken from Eady, 1968.

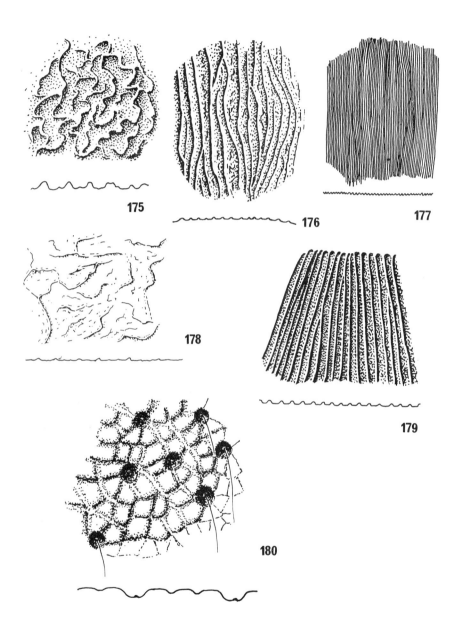

Figs 175–180. Cuticular microsculpture. 175, rugose. 176, strigose. 177, aciculate. 178, rugulose. 179, striate. 180, coriaceous, with punctures. The line below each figure represents an imaginary section of the cuticle. All figures are taken from Eady, 1968.

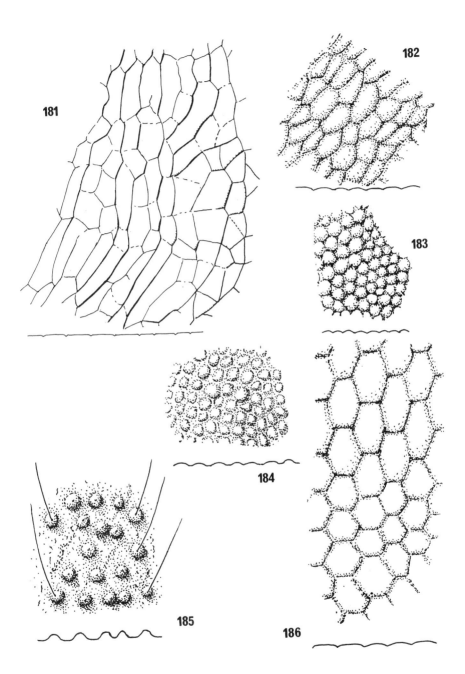

Figs 181–186. Cuticular microsculpture. 181, alutaceous. 182, coriaceous. 183, granulate. 184, pustulate. 185, papillate, with hairs. 186, reticulate-coriaceous. The line below each figure represents an imaginary section of the cuticle. All figures are taken from Eady, 1968.

Index

The index includes names of pimplines and hosts. Principal page references to the pimpline taxa are shown in bold. Synonyms and other invalid names are in italics. Illustrations are not indexed.

comariana 74
communis 87
compunctator 79
compunctor 21, 79, **83**
concors 19
conica 65
conicolana 55
coniferana 44
consociella 53, 74, 81
consortana 52
conspicuella 52
contemplator 20, 76, **78**
conturbatella 52, 54, 55, 74
corniger 89, 90
cornutus 58, 65, 66
coryli (Calocasia) 80
coryli (Phyllonorycter) 52
corylifoliella 52, 72
Corynopus 87
Cossidae 42
crassisetus 18, 41, **42**
crataegana 74
crataegi 80
cribrella 54
Croesia 53, 73, 74
cuculipennella 52, 54, 72
Cucullia 80
cucurbitina 58, 65
culiciformis 40, 42
culpator 18
Curculio 54
Curculionidae 38, 40, 52, 54, 55, 74, 90
curticauda 20, 70
curvipes 22, 91
cyanea 87
Cyclosa 65
Cydia 42, 44, 52, 54, 55, 81
Cynaeda 81
Cynthia 45

Dasychira 56
Dasysyrphus 78
daucella 54, 74
dealbana 74
deciduana 55
decorata 20
degener 20, **66**
Deilephila 80
Delomerista 17, 21, 24, 26, 84, **85**
Delomeristini 5, 21, **84**
Dendroctonus 40
dentalis 81
denticulella 52
deplanata 18
Depressaria 54, 74, 81
desvignesii 17, 21
detrita 18, **46**, 48
Deuteroxorides 21, 22, **90**
devoniella 52
Diachrysia 79

Diacritini 22, **91**
Diacritus 22, 23, **92**
Diadegma 54
diadematus 58, 65, 68
Dichrorampha 52
Dictyna 68
didyma 19, 55, **56**
diluta 18, **45**
diniana 73, 81
Dinocampus 74
discolor 20, **69**, 70
dispar (Lycaena) 54, 80, 84
dispar (Lymantria) 7
dissoluta 45
Ditula 74, 79
Diurnea 52, 72, 84
diversicostae 17, 37, **38**
divinator 21, **87**
Dolichomitus 17, 25, 26, 27, **36,** 41
dolichura 55
dorsalis 66
Dreisbachia 19, 25, 26, **62**
Drepana 72, 80
Drepanidae 72, 80

Ectropis 46
ekebladella 50
Elachistidae 52
elegans 18, 49, **53**
elevator 21, **90**
elongella 50, 72
Endothenia 52, 53
Endromopoda 16, 17, 18, 25, 27, **45,** 48
Ennomos 80
Enoplognatha 59
epeirae 19
Ephialtes 5, 17, 25, 27, **35,** 36, 74
Ephialtes 21, 83
Ephialtinae 17
Ephialtini 4, 5, 17, **35**
Ephialtini 20
Epiblema 52, 54, 78
Epinotia 52
Epirrita 81
Epiurus 18
eremita 89, 90
Eremochila 18, 44
Eriocrania 50
Eriocraniidae 50
Eriogaster 80
erraticum 59, 63
Eucosma 52, 73
eucosmidarum 18, 50, **54**
Eudemis 78
euphorbiae 55, 80
Eupithecia 52, 81
Euproctis 80, 81
Eurytomidae 46
Euura 52, 55
evonymella 81

Lasiocampidae 56, 74, 80
lateralis 36
Lathronympha 52
lautella 52
lecheana 74, 84
Leiopus 90
Lepthyphantes 66, 67
lethifer 87
leucapennella 54
Leucoma 74, 80
leucomela 87
libatrix 84
Limnaecia 81
linariata 52
linearis (Hartigia) 53, 54
linearis (Scambus) 19
lineolea 53
Linyphia 67
Liotryphon 18, 26, 27, 36, 40, **41**
Lipara 47, 48
Lissogaster 74
literosa 46
Lithosia 80
littoralis 52
Lixus 54
Lobesia 52, 74
loeflingiana 84
Lomaspilus 72
longana 74
longicollis 91
lucens 48
Luffia 81
lugubris 87
lunata 69
lutescens 63
Lycaena 54, 80, 84
Lycaenidae 54, 80, 81, 84
Lycia 42
Lymantria 7
Lymantriidae 56, 76, 80, 81

machaon 76, 80
Macrocentrus 52
maculator 8, 20, 71, **73**
maculicoxis 88
madida 20, 66, **67**
maestingella 52
Malacosoma 80
malvella 52
mandibularis 17, 21, 85, **86**
manifestator 17, **35**, 36
mediator 21, 87
Megarhyssa 91
melanacrias 21, 75, 76, **80**
melanocephala 20, 71, **74**
melanopyga 18
mengei (Lepthyphantes) 67
mengei (Metellina) 67
merianae 67
Mesites 38, 90

mesocentrus 17, 27, **38**
Mesoligia 46
messor 17, **38**
Metellina 67
Meteorus 73, 81
Metriostola 81
Metzneria 52, 53
metzneriella 52
micans 40
microdactyla 54
Microgaster 74
millenniana 55
Mimas 80
minor 90
Molorchus 90
Moma 80
Mompha 52, 54, 55, 74
Momphidae 42, 52, 54, 55, 74, 81
montana (Linyphia) 67
montana (Tetragnatha) 68
mordax 40
morosa 54
Morphophaga 53
moschata 90
multicolor 17, 19, **59**
Myelois 54
myllerana 74
myopaeformis 42
mystaceum 69

nebulosus 90
neglecta 63
Nemophora 53
nemoratus 54
Neosphaleroptera 72
Neoxorides 17, 88
Neoxoridini 21, 88
nervosa 52
neustria, 80
nicellii 52
nielseni 17, 19, 64, **65**
nigricana 52
nigricans 18, 45, 48, 49, 53, **54**
nigricornis 17, 20, **68**
nigricoxis 17, 18, 46, **47**
nigrinus 46
nitens 17
nitida 18, 46, **47**
Noctuidae 45, 46, 53, 55, 56, 74, 80, 84, 86
notata 17, 21, **89**
Notodontidae 56, 80
novita 21, 85, **86**
nubiferana 72
nubilana 72
nucifera 91
nucum 18

oblongus 58
obscurepunctella 52
obtusa 68

occidentis 74
ochropoda 54
ocularis 80
oculatoria 19, 57, **58**
Oecophoridae 52, 54, 72, 74, 78, 81, 84
Olethreutes 74, 80, 81
Omalus 87
Oncocera 81
onotrophes 53
opacellata 21
Operophtera 74, 78, 79
opistographa 65
Orchestes 52
Orellia 54
Orgyia 56, 72, 80, 81
ornata 19, 57, **58**
Orthosia 52
Osmia 87
ovata 59
ovivora 17, 19, **58**
Oxyrrhexis 17, 61

padella 72
paleana 54
pallipes 20, **68**
pallipes 67
Pammene 42
Pandemis 52, 74, 84
Papilio 76, 80
Papilionidae 76, 80
Paraperithous 18, 26, 27, **40**
Pararge 84
Parornix 52
Passaloecus 87, 89, 90
pastinacella 54, 74, 81
pavonia 80
paupella 52, 53
pectinicornis 89
pedunculi 55
peltata 67
Pemphredon 87
percontatoria 20, **69**
percontatoria 20
Peribatodes 80
Perilitus 74
Perithous 6, 8, 17, 21, 24, 26, 84, **86**
Perittia 52
persuasoria 22, 88, **91**
Petrova 44, 55, 87
Pexicopia 52
pfankuchi 17, 21, 85, **86**
Phidias 22
Philodromus 58
Philudoria 56, 74, 80
Phobocampe 72, 74
phoenicea 20, 70
Phragmatobia 73
phragmitella 81
phragmitidis (Arenostola) 45
phragmitidis (Endromopoda) 18, 46, **47**

Phyllonorycter 52, 55, 72
Phytodietus 52, 53, 74
pictifrons 19, **62**
Pieridae 80, 81, 84
Pieris 80, 81, 84
Pima 52
Pimpla 5, 7, 8, 9, 17, 20, 24, **74**
Pimpla 17, 35, 70, 71
Pimplini 5, 20, **70**
Pimplini 17
pini (Dendrolimus) 56
pini (Pissodes) 40
Piogaster 19, 23, **63**
pisi 80
Pissodes 40, 55
planatus 18, 49, **54**
Platyedra 52
Platyptilia 52
Pleuroptya 84
pneumonanthes 52
podagrica 19, **63**
Podoschistus 17, 22, **90**
Poemenia 17, 21, 22, **89**
Poemeniini 5, 21, **88**
Polemophthorus 20
Polysphincta 17, 19, 23, 62, **64**
Polysphinctini 5, 6, 8, 19, **61**
pomonella 42, 81
pomorum (Anthonomus) 55
pomorum (Scambus) 8, 18, 49, **55**
Pontania 55
populella 72, 74
populnea (Saperda) 38
populneus (Dolichomitus) 17, 37, **38**
porcellus 80
potatoria 80
pronubana 72
Pristiphora 55
processioniae 20
Prochoreutis 74
profundana 78
prolongata 91
Promecotheca 91
proxima 55
pruni 84
Psenulus 87, 88
Pseudopimpla 35
Pseudorhyssa 6, 8, 21, 22, 38, 84, **88**
Psyche 72, 81
Psychidae 72, 73, 74, 78, 81
pterelas 17, 37, **40**
Pterophoridae 52, 53, 54
Ptilinus 89
Ptocheuusa 52, 53
Ptycholoma 74, 84
pulchrator 20
punctator, 79, 80
punctulata (Piogaster) 19, **64**
punctulatus (Liotryphon) 18, 41, **42**
putridella 52